INSECT SCIENCE-DIVERSITY, CONSERVATION AND NUTRITION

Edited by **Mohammad Manjur Shah**
and **Umar Sharif**

Insect Science-Diversity, Conservation and Nutrition
http://dx.doi.org/10.5772/intechopen.71639
Edited by Mohammad Manjur Shah and Umar Sharif

Contributors

Claude Wicker-Thomas, Béatrice Denis, Benjamin Morel, Michael Samways, Alexa Alexander, Vagner Arnaut De Toledo, Emerson Dechechi Chambo, Regina Garcia, Carlos A Carvalho, Ludimilla Ronqui, Claudio Silva Júnior, Pedro Santos, Fernando Cunha, Daiane Oliveira, Maiara Caldas, Nardel Soares, Mudasir Ahmad Dar

Notice

Statements and opinions expressed in the chapters are these of the individual contributors and not necessarily those of the editors or publisher. No responsibility is accepted for the accuracy of information contained in the published chapters. The publisher assumes no responsibility for any damage or injury to persons or property arising out of the use of any materials, instructions, methods or ideas contained in the book.

First published in London, United Kingdom, 2018 by IntechOpen
IntechOpen is the global imprint of INTECHOPEN LIMITED, registered in England and Wales, registration number: 11086078, The Shard, 25th floor, 32 London Bridge Street
London, SE19SG – United Kingdom
Printed in Croatia

British Library Cataloguing-in-Publication Data
A catalogue record for this book is available from the British Library

Additional hard copies can be obtained from orders@intechopen.com

Insect Science-Diversity, Conservation and Nutrition, Edited by Mohammad Manjur Shah and Umar Sharif
p. cm.
Print ISBN 978-1-78923-454-1
Online ISBN 978-1-78923-455-8

Meet the editors

Dr. Mohammad Manjur Shah obtained his PhD degree from Aligarh Muslim University in 2003. He has been actively working on insect parasitic nematodes and has pioneer in this field regarding the north-east part of India. He has presented his findings at several conferences and published his articles in various reputed international journals. He completed post-doctoral fellowship twice under the Ministry of Science and Technology, Government of India before becoming the Senior Assistant Professor at the Northwest University, Kano, Nigeria, in 2015. In addition to this book, he has edited three books for IntechOpen. He is also a reviewer of several reputable international journals. He has been listed in various biographies published in USA and the UK.

Dr. Umar Sharif received his MSc and PhD degrees from Byero University, Kano, Nigeria. At present, he is the Head of Department of Biological Sciences, Yusuf Maitama Sule University. Since 1990, he has been actively involved in teaching as well as various research activities, in addition to his administrative responsibilities. His has experience in the field of pathology. He is a member of various scientific societies in Nigeria and has experience in constructive criticism and review in various fields of biology. He has presented his findings in various conferences and published papers in reputable international journals.

Contents

Preface

This book is mainly intended for biologists, in particular general biologists and entomologists. This book discusses the recent advances in basic and applied approaches including research on the genetics of insects, its application in resolving the consequences of world population growth, its impact on agriculture and control strategies and their implications on fast-depleting insect resources. The application of insects as a probable nutrient substitute along with the role of sex hormones among insects has been thoroughly discussed.

The book was created by collecting expert opinions in the field from different countries and we are fortunate to have included very interesting and important articles on some challenging areas of entomology. The entire book basically contains five chapters spread over two sections: Section I mainly focuses on diversity, conservation and nutrition, while Section II is concerned with economic importance and up-to-date information on the role of peptides. The book is well illustrated with diagrams, graphical representations and flow charts for easy understanding the important information discussed in the book.

The editor is very grateful to all the associated staff for making this book project successful. I hope this book will be useful to a wide range of readers by giving them a better understanding of the subject.

Dr. Mohammad Manjur Shah
Yusuf Maitama Sule University
Kano, Nigeria

Dr. Umar Sharif
Yusuf Maitama Sule University
Kano, Nigeria

Diversity, Conservation and Nutrition

Chapter 1

Diversity and Functions of Chromophores in Insects: A Review

Tanuja N. Bankar, Mudasir A. Dar and
Radhakrishna S. Pandit

Additional information is available at the end of the chapter

http://dx.doi.org/10.5772/intechopen.74480

Abstract

Insects are the most diverse among the animal kingdom. The diversity of insects is ever increasing due to their fast adaptability to the rapidly changing environmental conditions. The physiology of insects plays a vital role in the adaptation and competing adjustments in the nature with other species. The mechanism of vision and the involvement of visual pigments, like chromophores particularly in flies, have proved to be landmarks in the field of research. This has been achieved with the discovery of novel pathways involved in the mechanism of pigment development. However, certain visual pigments and their relationship with various chromophores need to be further elaborated. The role of insect pigments in vision, to identify the hosts, prays, and predators, is also discussed. Many naturally occurring pigments of insect origin are continuously being explored for better prospects and human welfare. The abundant availability of insect species all over the world and the never ending task of exploring their potential at morphological, physiological, evolutionary, and genetic levels have a tremendous potential to explore the subject like entomology.

Keywords: insects, chromophore, genetic, ommochrome, pigment, xanthommatin

1. Introduction

Insects represent one of the largest groups of animals on earth, which constitute over 1 million species and still counting (Gross 2006) serving many trophic roles like pests and pollinators in food chains. We are surrounded by a large variety of insects which always attract our attention with their intriguing beautiful color patterns. Pigmentation assists various species of insects in many biological activities, such as camouflage, mimicry, aposematism or warning, selection for sex, and communication by signaling [1]. Apart from this, other pigments produced by insects are involved in the metamorphosis, growth, and developmental stages of any colorful insect's life cycle. There are pigments and chromophores [2] that are known to play a vital role in imparting vision to insects for their routine activities. It is interesting and intriguing to know how these colors are formed in insect body. Here, we have addressed the questions related to chromophores found in the eye pigment of insects along with other visual eye pigments.

2. Synthesis of pigments in insects

In insects, the epidermis produces pigments via a series of developmental stages. This pattern formation and synthesis of pigments influence the phenotypes and behavior of insects in one or other way. Most of the insect pigments are either synthesized by insects such as, anthraquinines, pterins, tetrapyrroles, ommochromes, and papiliochromes, or absorbed from the antioxidative carotenoids and flavonoids of their host plants [3]. Apart from imparting body coloration, ommochromes act as visual pigments, melanins protect against ultraviolet radiation, and tetrapyrroles facilitate oxygen transport to cells. Insect pigmentation has been studied in detail by Mollon in most common insect model, *Drosophila melanogaster* [4]. The process of pigmentation in insects occurs in two stages, viz., location or appearance of pigments in space or time and the biological as well as chemical synthesis. These processes are controlled by patterning genes which regulate the distribution of pigments and their effector genes. For instance, in butterfly, *Bicyclus anynana*, a protein called "Engrailed" synthesizes a yellow pigment in later stages which forms the golden ring adult eyespots on the hind wing. While some researchers have revealed that effector genes are responsible for enzymatic pigment production.

2.1. Insect pigmentation: genetics and evolution

Study of pigmentation system is vital for study of links of genetic changes to the evolutionary variation in fitness-related traits of insects [5]. For example, pigmentation of the normal eyes is known to be blocked by majority of mutations in ocelli during the synthesis of brown eye pigment xanthommatin [6]. Mason and Mason [7] have reviewed the current state of comparative biology in context to pigmentation. Advanced studies carried out on genetic analysis of pigmentation in lower vertebrates, mice, and humans elucidate various aspects of development and evolution of the process of pigmentation at different stages. Molecular studies in lower vertebrate pigmentation and a comparative account of genes in different species arising from a common ancestral gene have been fruitful in the study of pigmentation in various insect species [8]. Evolutionary studies between mammals and other vertebrates have revealed significant differences in pigmentation mechanisms between these species. Such data provides an overall view of pigments and their existence across numerous species [7]. Briscoe and Chittka [9] have reviewed the physiological and molecular mechanisms of insect color vision. Recently, role of a marker which is dominantly expressed during insect transgenesis has been elucidated by Takahiro et al. [10].

2.2. Use of pigment in insect

Carotenoids are uniquely involved in functional dynamics of almost all green-colored insects. Heath et al. [11] reviewed the various carotenoids and their derivatives for function and influence of their interactions between their environments, such as vegetation on which they thrive. They also reviewed the biological synthesis as well as structure of these compounds and discussed their roles in various phenomena, such as warning coloration, vision, photoperiodism, and diapause, along with their antioxidative role in signaling. Further, they explored the probable functions of carotenoid derivatives such as strigolactones and apocarotenoids in mediating interactions between insects, plants, fungi, and their parasitoid enemies [11]. *Manduca sexta* larvae appear blue and green in color when fed on artificial and natural diets (green plants), respectively. The green pigment is made up of two chemicals, namely, biliverdin, which is a blue pigment, and lutein, a yellow pigment. Artificial diet has very little lutein, and *Manduca* spp. are unable to synthesize lutein on their own and have more of a blue color than their plant fed counterparts. In *Manduca sexta*, lutein is the only carotenoid absorbed from the diet. This is because special transporters in their gut responsible for absorption of carotenoids recognize only lutein. Strong chemical reagents modify the colors of lepidopterous pigments or in some cases dissolve them out of the wings [12].

2.3. The role of pigmentation in insect vision

There are three types of eyes found in insects, namely, simple, apposition compound, and superposition compound eyes, as shown in **Figure 1**. The color vision of insects can discriminate wavelengths in varying ranges. Honey bee is the best example of this phenomenon.

Adult insects naturally have three simple eyes on the top of their heads which are made up of a lens and an extended retina. Some dorsal ocelli having either tapeta or a mobile iris can view at an angle of 150° or more and as many as 10,000 receptors like those of hunting spiders are present in a single eye. They become more or less out of focus in a condition where the retina and lens are close to each other. Ocelli are horizon detectors which control the response of receptors to variations of intensity, and distance of light perceived by the insect from the environment and contributes to flight equilibrium. The dorsal ocelli of adult locust have a very typical arrangement (**Figure 2**).

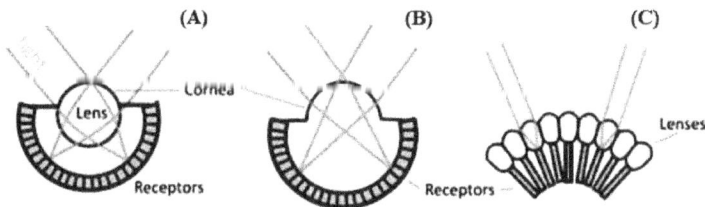

Figure 1. Types of insect eyes (A) simple eye, (B) apposition compound eye, (C) superposition compound eye. (*Source:* [65]).

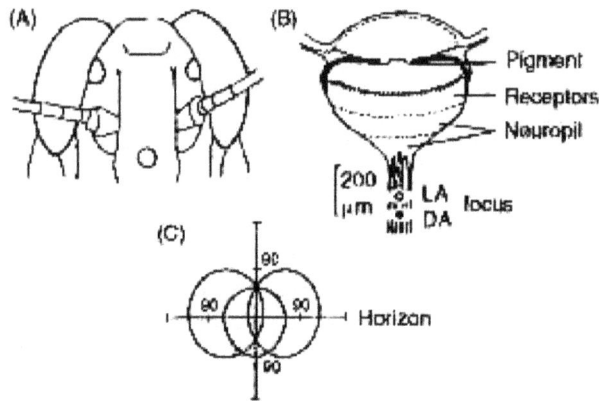

Figure 2. Dorsal ocelli of adult locust: (A) frontal and lateral ocelli on head and (B) section of an ocellus with different layers and positions of the long distance focus behind the receptor tiers in the presence and absence of light. (C) Skyline view of the three ocelli (*source*: http://what-when-how.com/insects/eyes-and-vision-insects/).

The spectral sensitivity in receptors of eyes of honeybees suggests that the visual pigments in insects are rhodopsins, consisting of protein bound to the retina of the eye [13]. Some Lepidopteran insects have color vision with spectral sensitivity suggesting that the eyes contain two photopigments. The photoreceptors form a large part of the eye in sphingid moth, *Deilephila elpenor*, and most of the screening pigments can be separated out from the receptors [14]. Chemical nature of color vision depends upon the light absorption from the external sources in form of electromagnetic radiation. Typical pigments in the eye act as transducers and convert electromagnetic energy into the chemical energy. This stimulates an impulse within a nerve cell or neuron. All insect visual pigments are produced by retinula cells and stored in the rhabdoms of the compound eyes and ocelli. Only two types of visual pigments, one pigment absorbing green and yellow light and the other absorbing blue and ultraviolet (UV) light, are predominant in insects. However, red color is invisible to insects. Insects have limited color vision only when the frequency of response lies within the UV range. Bichromatic insects having two types of pigment receptors fail to distinguish between single colors and mixture of colors. Trichromatic insects such as honeybees, bumblebees, and most diurnal butterflies possess three types of receptors that are known to have true color vision. It means that they can perceive a complete spectrum of colors and also discriminate between individual colors and color mixtures. Eyes of trichromatic insects have three visual pigments having absorption maxima at UV (360 nm), blue-violet (440 nm), and yellow (588 nm) wave lengths. In any bichromatic insect, both types of receptors are stimulated due to which a combination of UV with yellow, which are at the extremes of insect's visual spectrum, appears as blue green. However, the same combination appears as two separate colors to a trichromatic insect since the receptor of blue-violet remains unstimulated. Bees perceive the unique color combination of UV yellow as equivalent to the purple in humans. Thus, it is the "bee-purple" in bee's color vision. All the peripheral rhabdomeres in ommatidia of *Notonecta glauca* contain a visual pigment which is sensitive to red color having wavelength of greater than 700 nm. In *Notonecta glauca*, on the dorsal region of the eye, both rhabdomeres in a single ommatidium contain either a pigment with absorption maxima at 345 nm or absorption at 445 nm in adjacent rhabdoms. In the ventral part of the eye, central rhabdomeres contain a pigment having maximum absorption in UV range. Variations in spectral absorption in various types of screening pigments were also studied by Schwind et al. [15]. Shozo [16] studied the effect of different intensities of light on the visual pigments and their adaptive evolution. The authors elucidated that critical amino acids involved in spectral orchestration and their interactive effects on spectral shifts are necessary for the molecular function [16].

2.4. Insect eye pigments: pteridines and xanthommatin

Moraes and coworkers [17] were the first to carry out spectral studies for black and red pigment color absorption by insect eye and later compare them with that of *D. melanogaster*. Both the black- and red-type eye forms of *Triatoma infestans* are made up of ommochromes of the xanthommatin type (**Figure 3**). However, eye pigments, namely, pteridines, melanins, and

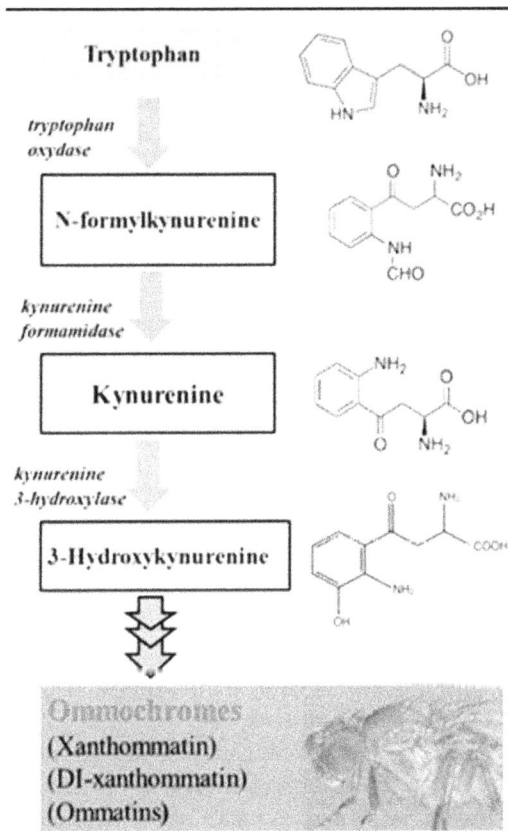

Figure 3. A generalized ommochrome pathway in insects.

ommins, were absent in *T. infestans* unlike *D. melanogaster*. This variation in color of the eye was due to activity of the xanthommatin concentration [17, 18]. Pigments extracted from eyes of wild-type mosquito *Anopheles gambiae* revealed the presence of the ommatin precursor 3-hydroxykynurenine, its transamination derivative xanthurenic acid, and a dark, red-brown pigment spot that probably is composed of two or more low-mobility xanthommatins. No colored or fluorescent pteridines were evident. Some insect colors are a result of mutations that occur in insects during their developmental stages. Mosquitoes homozygous for an autosomal recessive mutation at the red-eye (r) locus have a brick-red-eye color in larvae, pupae, and young adults, in contrast to the almost black color of the wild eye. Mosquitoes homozygous for this mutant allele have levels of ommochrome precursors that are non-distinguishable from the wild type, but the low-mobility xanthommatin spot is ochre-brown in color rather than red-brown as in the wild type. Mosquitoes with two different mutant alleles at the X-linked pink-eye (p) locus, which confers a pink-eye color, and a white eye phenotype (pw) in homozygotes or hemizygous males have normal levels of ommochrome precursors but no detectable xanthommatins. Mosquitoes homozygous for both the *r* and *p* mutant alleles have apricot-colored eyes and show no detectable xanthommatins. Both the pink-eye and red-eye mutations involve defects in transport or assembly of pigments in the membrane-bound pigment granules rather than defects in ommochrome synthesis [19]. Ferré et al. [20] analyzed the contents of pigments xanthommatin and dihydro-xanthommatin which are responsible for causing brown eye color and related metabolites' "garnet" gene in the eye color mutants of *D. melanogaster*. Pteridines the fluorescent metabolites of the xanthommatin pathway responsible for red-eye color were also quantitated. The authors concluded that the synthesis and accumulation of xanthommatin in eyes may be related to pteridine synthesis pathway give rise to isoxanthopterin, drosopterins, and biopterin as final products. Beard et al. [19] reported that pigments in eyes of wild-type mosquito, *Anopheles gambiae*, show the presence of the ommatin precursor (3-hydroxykynurenine) and a dark red-brown pigment spot composed of two or more low-mobility xanthommatins. Pteridines, however, were found to be absent. They showed the color variations seen in the mutants and compared them with the wild-type eye color. The pink and the red-eye mutations involved defects in the transport into assembly of pigments in the membrane-bound pigment granules rather than defects in ommochrome synthesis [21].

3. Screening pigments and rhodopsins

Pigment cells contain screening pigments which are determinants of eye color in insects. The red-colored screening pigments of eyes of the fly permit stray light to photochemically restore photo-converted visual pigments. Many insect species have dark-colored eyes, with distinctly featured color patterns. A large variety of flies and butterflies were studied by the pioneer Entomologist Stavenga, D. G., to bring forth physical and functional aspects of eye colors in insect color vision [22]. The yellow pigment granules located in photoreceptor cells act like the pupils of the eye which control the light sensitivity and adjust it accordingly. The eyes of most insect possess black screening pigments which protect the photoreceptors from stray light entering the eyes of insect. Eyes of tabanid flies are strongly metallic in color, due to the multilayered cornea. Such corneal patterns are seen in golden green eyes of deer fly, *Chrysops relictus*. The sensitivity spectrum of photoreceptors with green absorbing rhodopsin is narrowed due to reduction in orange green color transmission of the corneal lens. In contrast, the spectral sensitivity of proximal long wavelength photoreceptors is enhanced by the tapetum in eyes of butterflies and regularized by the pigment granules lining the rhabdom [55]. Kim and colleagues [24] reviewed the peculiarities of the *Drosophila* spp. red-eye pigments and their genes and enzymes involved in its biosynthetic pathway. The retina of the adult tobacco hornworm moth *Manduca sexta* contains three visual pigments, namely, a green sensitive rhodopsin and smaller amounts of blue sensitive and ultraviolet sensitive rhodopsins. Similarly, White and coworkers [63], studied seventeen stages, each stage representing of one to two days of chronological age in the morphological system of *Manduca sexta*. Progressive maturation of the retina in ultrasections was monitored to measure rhodopsin in sections of the retina, and electroretinograms were recorded from stages 8–17.

3.1. Papiliochromes and pterins

Papiliochromes and pterins both present in the ommatidia of eyes in insects are synthesized from amino acids tyrosine, tryptophan, and guanosine triphosphate (GTP), respectively. Chemochromes make the insects attractive by providing striking colors to their appearance and provide functional benefits for the commercially important insects [3].

3.2. Visual pigments of the fruit fly

The vast amount of information available on the fly visual system provides a detailed information regarding other insect species [8]. When light quanta hit upon the visual pigment molecules of the eye, any insect is able to visualize. Hamdorf [26] measured the number of microvilli in the rhabdomere of a blowfly and concluded that a microvillus contains well over 1000 visual pigment molecules, such that, a photoreceptor is made up of approximately 2×10^8 visual pigment molecules [26]. Stavenga and Smakman [27] measured the visual pigment content of Blowfly RI6 photoreceptors in order to determine their spectrum and polarization sensitivities within a particular wavelength range. It is seen in flies that self-screening increases the spectral sensitivity whereas relative UV sensitivity is lowered when the visual pigment content is high. However the electrical response remains unaffected by the amount of visual pigment [27].

4. Synthesis and renewal of visual pigment

The protein opsin which binds to its extraordinary chromophore 3-hydroxyretinal is abundantly present in the visual pigment of the fruit fly [28]. When the rhodopsin state is achieved, the chromophore preexisting in the *cis* configuration gets converted into *trans* isomer after photon absorption. In the next thermostable metarhodopsin state, thermal decay occurs at different stages in a stepwise manner. Reconversion to rhodopsin occurs in the RI6 principal cell metarhodopsin molecule in houseflies [29] also. Stavenga et al. [30] found a strong variation in peak wavelengths of rhodopsin and metarhodopsin in blowfly and hoverflies, e.g., *Eristalis* spp. [30]. Photopigments of all animals are composed of a large membrane-spanning protein, the opsin which enfolds the aldehyde of vitamin A, as the chromophore [26, 30, 31]. Chromophores are produced in the presence of retinoids during the process of visual pigment synthesis. Release of chromophore after the intracellular breakdown of opsin and its transport from the visual sense cell to the primary pigment cell in the presence of a light driven isomerase was studied by Schwemer [32]. For the purpose of compass orientation, most insects depend upon the sky polarization pattern, and some insects make use of the sky chromatic contrast [33]. They also identified an opsin of a UV-absorbing visual pigment and studied expression of DRA receptors. The retina of insect eye contains two or more types of cones containing photo pigments with different spectral sensitivities [34]. Light of most visible wavelengths produces a unique pattern of activity among the different cone types. These patterns encode the distribution of wavelengths across the retinal image. When light quanta are absorbed by the chromophores, they change shape and activate the opsin, which then functions as a catalyst for further reactions in the photo receptor. The spectral sensitivity of the photopigment is that of the chromophore, modified by the proximity of the opsin, which disturbs the arrangement of the vitamin A aldehyde. Changes in one amino acid group at critical points in the opsin can significantly alter the spectral sensitivity of the opsin chromophore combination [35]. Because the amino acid sequence of the opsin is determined genetically, mutations through evolution have produced a wide range of photopigments with spectral sensitivities often matched to the ecological niches of the animals [4, 31, 36, 37]. Insects have also developed other strategies for shaping the spectral sensitivities of their ommatidial photoreceptors. Alternating layers of material in the corneal facet can serve as interference filters, permitting a restricted range of wavelengths to reach the photopigment. Neighboring ommatidia can be adapted to different parts of the spectrum by varying the filtering properties of the cornea. The photo receptors of certain flies contain additional pigments which serve as "antennas" facilitating the capture of ultraviolet light and transmitting the energy to rhodopsin to initiate the visual process [38].

4.1. Photoreceptor cell

Goldsmith et al. [39] studied the effect of diet on production of visual pigment. They concluded that carotenoid replacement due to the presence or absence of Vitamin A, promoting production of visual pigment along with chromophore and opsin increment. Providing a deficient chromophore to *Drosophila* spp. and *Calliphora* spp. which are deprived of carotenoid leads to maturation of opsin due to the posttranslationally controlled expression of apoprotein [4, 40–42]. Arikawa et al. [43] demonstrated that there exist distal photoreceptors in the retina of butterflies belonging to genus *Papilio* which exhibits specific spectral sensitivities. A short-wavelength sensitive receptor exists, which may be a UV receptor (normal spectrum), a violet receptor (very narrow spectral bandwidth), a blue receptor, or a green receptor (double peak/single peak). Ommatidia contain only the violet receptor and single peak receptor, which are capable of emitting fluorescence in the presence of UV light. The fluorescence is emitted from a pigment which is located at the extreme end of the ommatidium which absorbs UV light, leading to narrow spectral sensitivity of violet receptor and a single peak green receptor [43]. A simple and accurate method was used for measurement of absorbance changes during saturating adaptations of the visual pigment to various monochromatic lights which was based on measurements of difference in spectral amplitudes. The predominant pigment absorbs maximally at rhodopsin and metarhodopsin wavelength [44]. Meinecke and Langer [45] reported that in the noctuid moth *Spodoptera exempta*, each ommatidium regularly contains eight receptor cells belonging to three morphological types: one distal, six medial, and one basal cell for different visual pigments within the eye. Langer et al. [46] also identified three types of visual pigments and their localization in the photoreceptor cells of compound eye of silk moth *Antheraea polyphemus* [46].

4.2. Sensitizing pigment

In flies, visual pigments bind to 3-OH-retinol along with the chromophore 3-OH- retinal [29]. As a result when flies are fed on diet deficient in vitamin A, they demonstrate a low visual sensitivity and show a declined UV sensitivity relative to the blue green peak [47]. The 3-OH-retinol or the sensitizing (antenna pigment) studied by Kirschfeld and his colleagues [48] absorbs UV and upon excitation by a photon transfers the absorbed energy to the chromophore which is later isomerized [48]. Energy transfer occurs from the excited sensitizing pigment to rhodopsin as well as metarhodopsin [49]. A strong enhanced spectral sensitivity in UV spectrum is caused due to the rhodopsin being sensitized by the UV-absorbing antenna pigment [50].

Hamdorf et al. [51] elucidated the in vivo electrophysiological aspect of the rate at which the retinoids get incorporated in the various visual pigments [51]. Role of retinoids in retinal degeneration in *Drosophila* mutant when exposed to various chemicals has also been studied [42]. Minke and Kirschfeld [49] demonstrated that a pigment which is photostable acts as a sensitizer for rhodopsin, and they used membrane potential to measure variations in visual pigment in fruit fly.

4.3. Chromophores

Hamdorf [26] reported behavioral and electrophysiological experiments in honeybee eyes as well as in neuropteran, *Ascalaphus macaronius* [26, 52]. Kashiyama et al. [53] carried out the molecular characterization of visual pigments in *Brachiopoda* spp. and showed that ancestors of *Pancrustacea* spp. and the insect, *Branchiopoda* spp., lineages possessed minimum of five or six types of opsins [53]. Further, Helmut et al. [54] reported the presence of three varieties of visual pigments in the retinal extracts of moth *Antheraea polyphemus*. In many *Drosophila* species, a genetic model for characterizing retinoid-binding proteins was established. It was reported by Tao et al. [55] that PINTA is expressed and is functionally required after the production of vitamin A in the retinal pigment epithelia. It was the first genetic evidence for the retinal pigment cells in visual response in *Drosophila* spp. [55].

4.4. Ommochrome pathway in insects

Ommochromes are biological visual pigments occurring in the eyes of crustaceans and insects, which determine the color of insect eye. Mostly, these are predominantly found in chromophores of cephalopods and spiders. Ommochromes are in the form of pigment granule deposits inside the cells of the hypodermis, just below the cuticle [56]. They are responsible for a wide variety of colors, ranging from yellow, red, and brown to black. Ommatins impart light colors, while combinations of ommatin and ommins are known to impart dark colors [56, 57]. In few insect lineages, ommochromes have special function of coloration of integument and tryptophan secretion. Only in family Nymphalidae, ommochromes are well known as butterfly wing pigments. In order to understand the occurrence of subcellular process during evolution, the development of pigment ommochrome called xanthommatin in the wings of nymphalid butterfly *Vanessa cardui* was identified and explored. Fragments of ommochrome enzyme genes, "Vermilion" and "Cinnabar," were cloned with the well-known precursor transporter gene called "White." These genes were found to have transcribed at high levels during the development of the wing scale tissue. However, the transcription pattern and adult pigment patterns were not associated with each other. These results indicate that there exists a transcriptional interrelationship between pre-pattern and pigment synthesis in *Vanessa cardui* [58]. The color of eye shine of some butterflies is determined mainly by the reflectance spectrum of the tapetal mirror and the transmittance spectrum of the photoreceptor screening pigments [23]. Insausti et al. [59] studied the morphological and physiological changes associated with mutation in the red-eyed mutant bug, *Triatomine* sp. They demonstrated ommochromes as one of the major pigments responsible for coloration of eggs, eyes, and body surface of insects [59]. However, final steps of molecular mechanisms of ommochrome pigment synthesis are not known. Osanai-Futahashi et al. [60] identified the gene involved in egg or eye pigmentation, and it has been identified and characterized in *Bombyx* and *Tribolium* species [60] as well.

5. Discussion

Evolutionary questions about process of pigmentation highlight the similarities and differences between various organisms in a framework. Thus, developmental and evolutionary data is useful for creation of a unified view of insect pigment cells and to study its existence across diverse species [7]. Briscoe and Chittka [9] reviewed the physiology, molecular biology, and neural mechanisms of color vision of insects. Studies on phylogeny and analysis at molecular level revealed that the basic bauplan, UV-blue green trichromacy dates back to the Devonian ancestor of all pterygote insects. In addition to exploring these factors, quantification of variance between individual and population of insects and fitness measurements was used to test the adaptiveness of characteristics in insect color systems [9]. The molecular basis of spectrum analyses in vision pigments can be elucidated by conducting experiments to study the adaptation of different insect species to various light conditions with time. To explain the molecular and functional aspect of visual pigment adaptations in a better way, it is necessary to understand all important molecule exchanges that may be involved in the alignment of spectra and investigate effectiveness of the interactions of spectral shifts [16]. A number of examples of fly and butterfly species possessing dark-colored eyes are known, but distinct colors or patterns are discussed to depict current knowledge available on the physical and functional implications in insect ocular color [61]. Color vision has its greatest value for species that are active during the day when there is abundant light to illuminate objects with different spectral reflectance. Thus, color vision is particularly well developed in various species such as birds, reptiles, and some fishes which trace their evolution through long lines of diurnal ancestors. Humming birds, as well as chickens and pigeons, may have as many as four different cone photo pigments, allowing them to make fine color discriminations over a wide range of wavelengths. Avian cones, like those of certain turtles and amphibians, also contain colored oil droplets which may further refine their spectral selectivity. The oil droplet can act as a filter, limiting the wavelengths that reach the photo pigment. In principle, one can possibly construct a color vision system with one photo pigment and different kinds of oil droplets in different cones, although this strategy does not appear to have been adopted in evolution. The functional significance of the cone oil droplets in birds, turtles, and amphibia still remains unknown [62]. Bernard et al. [63] described the red color absorbing visual pigment of butterflies. Cromartie [64] surveyed the knowledge about chemical nature and biogenesis of the coloring matters of insects. Importantly, the biological significance of important pigments occurring in insects has been mentioned by emphasizing on the remarkable developments [64].

6. Conclusions

Insects are tiny creatures having typical eye features which help them to visualize the world around them and unique pigments which impart beautiful colors to their body parts. The role of color in genetic evolution of many insect species has been studied in the past. The pigments which play a major role in the coloration, especially the pigments that are vital for the vision in insects, have been studied. However, further insight into the same is needed with the use of advanced techniques. The induced and spontaneous mutations related to pigmentation have been investigated in many insect species. The pattern of pigment synthesis and the stages involved during metamorphosis have also been elucidated. The mechanism of vision and the involvement of visual pigments, especially in flies, have proved to be a landmark in the field of research. This has been done by discovery of novel pathways and their detailed studies. However, certain visual pigments and their relationship with various chromophores need elaborate studies to be carried out. The role of insect pigments in vision with respect to identification of hosts and prey-predator interactions for identifying preys is an interesting area of future research. Naturally occurring pigments from insects are being explored for better prospects and welfare of mankind. Their varied applications in areas as edible colors and rich source proteins in food industry can be a subject of future research. The abundant availability of insect all over the world and the never ending task of exploring their potential at the morphological, physiological, evolutionary, and genetic level open up new avenues for a wide and more interesting subject of entomology.

Acknowledgements

Authors acknowledge the authorities of Savitribai Phule Pune University for providing the necessary infrastructural support.

Author details

Tanuja N. Bankar, Mudasir A. Dar and Radhakrishna S. Pandit*

*Address all correspondence to: rspandit@unipune.ac.in

Department of Zoology, Savitribai Phule Pune University, Ganeshkhind, Pune, Maharashtra, India

References

[1] Alcock J. Animal Behavior: An Evolutionary Approach, 6th ed. Sunderland, MA: Sinauer Associates, Inc.; 1998

[2] Kirschfeld K. Activation of visual pigment: Chromophore structure and function. In: The Molecular Mechanism of Photoreception. Series Dahlem Workshop Reports. Berlin, Heidelberg: Springer; 1986;**34**:31-49

[3] Gulsaz S, Ranjan SK, Pandey DM, Ranganathan R. Biochemistry and biosynthesis of insect pigments. European Journal of Entomology. 2014;**111**:149164. DOI: 10.14411/eje.2014.021

[4] Mollon JD. Colour vision: Opsin and options. Proceedings of the National Academy of Sciences of the United States of America. 1999;**96**:4743-4745

[5] Hoekstra HE. Genetics, development and evolution of adaptive pigmentation in vertebrates. Heredity. 2006;**97**:222-234. DOI: 10.1038/sj.hdy.6800861

[6] Vogt K, Kirschfeld K. Chemical identity of the chromophores of fly visual pigment. Die Naturwissenschaften. 1984;**71**:211-211

[7] Mason KA, Frost Mason SK. Evolution and development of pigment cells: At the crossroads of the discipline. Pigment Cell Research. 2000;**13**:150-155

[8] Stavenga DG, Schwemer J. Visual pigments of invertebrates. In: Ali MA, editor. Photoreception and Vision in Invertebrates. New York: Plenum; 1984

[9] Briscoe AD, Chittka L. The evolution of color vision in insects. Annual Review of Entomology. 2001;**46**:471-510

[10] Tao W, Craig M. Cellular/molecular rhodopsin formation in drosophila is dependent on the PINTA retinoid-binding protein. The Journal of Neuroscience. 2005;**25**:5187-5194

[11] Heath JJ, Cipollini D, Stireman JO III. The role of carotenoids and their derivatives in mediating interactions between insects and their environment. Arthropod Plant Interactions. 2013;**7**:1-20

[12] Meldola R. Pigments of Lepidoptera. Nature. 1982;**45**(1174):605-606

[13] Bernard GD, Remington CL. Color vision in 256 D.G. Lycaena butterflies: Spectral tuning of receptor arrays in relation to behavioral ecology. Proceedings of the National Academy of Sciences of the United States of America. 1991;**88**:2783-2787

[14] Höglund G, Hamdorf K, Langer H, Paulsen R, Schwemer J. The photopigments in an insect retina. In: Langer H, editor. Biochemistry and Physiology of Visual Pigments. Berlin Heidelberg New York: Springer; 1973. pp. 167-174

[15] Shozo Y. Evolution of dim-light and colour vision pigments. Annual Review of Genomics and Human Genetics. 2008;**9**:259-282. DOI: 10.1146/annurev.genom.9.081307.164228

[16] Stark WS, Ivanyshyn AM, Greenberg RM. Sensitivity of photopigments of R 1-6, a two-peaked photoreceptor in drosophila, Calliphora and Musca. Journal of Comparative Physiology. 1977;**121**:289-305

[17] Moraes AS, Pimentel ER, Rodrigues VL, Mello ML. Eye pigments of the bloodsucking insect, *Triatoma infestans* Klug (Hemiptera, Reduviidae). Brazilian Journal of Biology. 2005; **65**:477-481

[18] Muri RB, Jones GJ. Microspectrophotometry of single rhabdoms in the retina of the honeybee drone (*Apis mellifera*). Journal of Insect Physiology. 2014;**61**:58-65

[19] Beard CB, Benedict MQ, Primus JP, Finnerty V, Collins FH. Eye pigments in wild-type and eye-colour mutant strains of the African malaria vector *Anopheles gambiae*. Journal of Comparative Physiology. A, Neuroethology, Sensory, Neural, and Behavioral Physiology. 2002;**188**:337-348

[20] Ferré J, Silva FJ, Real MD, Ménsua JL. Pigment patterns in mutants affecting the biosynthesis of pteridines and xanthommatin in *Drosophila melanogaster*. Biochemical Genetics. 1986;**24**:545-569. PMID: 3092804

[21] Beard CB, Benedict MQ, Primus JP, Finnerty V, Collins FH. Eye pigments in wild- type and eye-color mutant strains of the African malaria vector Anopheles Gambiae. The Journal of Heredity. 1995;**86**:375-380. PubMed PMID: 7560874

[22] Stavenga DG. Color in the eyes of insects. IUBMB Life. 2013;**65**(4):334-340. DOI: 10.1002/iub.1145. Epub 2013 Feb 23

[23] Stavenga DG. Color in the eyes of insects. Journal of Comparative Physiology. A, Neuroethology, Sensory, Neural, and Behavioral Physiology. 2002;**188**:337-348. Epub 2002 Apr 13

[24] Kim H, Kim K, Yim J. Biosynthesis of drosopterins, the red eye pigments of *Drosophila melanogaster*. IUBMB Life. 2013;**65**:334-340. DOI: 10.1002/iub.1145

[25] Winderickx J, Lindsey DT, Sannocki EDY, Teller DY, Motulsky AG, Deeb SS. Polymorphism in red photopigment underlies variation in color matching. Nature. 1992;**356**:431-433

[26] Hamdorf K. The physiology of invertebrate visual pigments. In: Autrum H, editor. Handbook of Sensory Physiology VII/6A. Berlin: Springer; 1979. pp. 145-224

[27] Stavenga DG. Reflections on colourful ommatidia of butterfly eyes. The Journal of Experimental Biology. 2002;**205**:1077-1085

[28] Vogt K. Distribution of insect visual chromophores: Functional and phylogenetic aspects. In: Stavenga DG, Hardie RC, editors. Facets of Vision. Berlin, Heidelberg: Springer; 1989. pp. 134-151

[29] White RH, Brown PK, Hurley AK, Bennett RR. Rhodopsins, retinula cell ultrastructure, and receptor potentials in the developing pupal eye of the moth Manduca sexta. Journal of Comparative Physiology. 1983;**150**:153-163

[30] Stavenga DG, Smakman JGJ. Spectral sensitivity of blowfly photoreceptors: Dependence on waveguide effects and pigment concentration. Vision Research. 1986;**26**:1019-1025

[31] Jacobs GH. Cone pigments and color vision polymorphism. A comparative perspective in frontiers of visual science. Comparative Colour Vision. Washington, DC: National Academy Press. 1987;**176**:129-144

[32] Schwind R, Schlecht P, Langer H. Micro spectrophotometric characterization and localization of three visual pigments in the compound eye of Notonecta glauca L. (Heteroptera). Journal of Comparative Physiology. 1984;**154**:341-346

[33] Fabian S, Wakakuwa M, Tegtmeier J, Kinoshita M, Bockhorst T, Arikawa K, Homberg U. Opsin expression, physiological characterization and identification of photoreceptor cells in the dorsal rim area and main retina of the desert locust, *Schistocerca gregaria*. Journal of Experimental Biology. 2014;**217**:3557-3568. DOI: 10.1242/jeb.108514

[34] Schwemer J, Spengler E. Opsin synthesis in blowfly photoreceptors is controlled by an 11-cis retinoid. In: Rigaud JL, editor. Structures and Functions of Retinal Proteins. 1992. pp. 277-280

[35] Wittkopp PJ, Beldade P. Development and evolution of insect pigmentation: Genetic mechanisms and the potential consequences of pleiotrophy. Review Seminars in Cell and Development Biology. 2009;**20**:65-71

[36] Goldsmith T. Optimization, constraint and history in the evolution of eyes. Quart. Rev. Biol. 1990;**65**:281-322

[37] Osanai-Futahashi M, Tatematsu K, Yamamoto K, Narukawa J, Uchino K, Kayukawa T, Shinoda T, Yutaka B, Tamura T, Sezutsu H. Identification of the *Bombyx* red egg gene reveals the involvement of a novel transporter family gene in the late steps of the insect

ommochrome biosynthesis pathway. Membrane Biology: Journal of Biological Chemistry. 2012. published online April 2, 2012:**287**

[38] Menzel R. Spectral sensitivity and colour vision in invertebrates. In: Autrum H, editor. Handbook of Sensory Physiology VII/6A. Berlin: Springer; 1979. pp. 501-580

[39] Goldsmith TH, Barker RJ , Cohen CE. Sensitivity of visual receptors of carotenoid-depleted flies:A vitamin A deficiency in an invertebrate. Science. 1964;**146**:665-667

[40] Isono K, Tanimura T, Oda Y, Tsukahara Y. Dependency on light and vitamin a derivates of the biogenesis of insect retinal pigments 3-hydroxyretinal and visual pigment in the compound eye of *Drosophila melanogaster*. The Journal of General Physiology. 1988;**92**:587-600

[41] Reed RD, Nagy LM. Evolutionary redeployment of a biosynthetic module: Expression of eye pigment genes vermilion, cinnabar, and white in butterfly wing development. Evol. Dev. 2005;**7**(4):301-311

[42] Schwemer J. Visual pigment renewal and the cycle of the chromophore in the compound eye in the blowfly. In: Wiese K, Gribakin EG, Popov AV, Renninger G, editors. Sensory Systems of Arthropods. 1993. pp. 54-68

[43] Arikawa K, Mizuno S, Scholten DGW, Kinoshita M, Seki T, Kitamoto J, Stavenga DG. An ultraviolet absorbing pigment causes a narrow-band violet receptor and a single peaked green receptor in the eye of the butterfly *Papilio*. Vision Research. 1999;**39**:1-8

[44] Nicol JAC. Studies on the eyes of fishes. Vision in Fishes: NATO Advanced Study Institute Series. 1975;**1**:579-607

[45] Meinecke CC, Langer H. Localization of visual pigments within rhabdoms of the compound eye of *Spodoptera exempta* (Insecta: Noctuidae). Cell and Tissue Research. 1984;**238**: 359-368

[46] Langer H, Schmeinck G, Friedericke A. Identification and localization of visual pigments in the retina of the moth, *Antheraea polyphemus* (Insecta Saturnidae). Cell and Tissue Research. 1986;**245**:81-89

[47] Stavenga DG. Insect retinal pigments: Spectral characteristics and physiological functions. Pigment of Eye. 1982;**11**:231-255

[48] Kirschfeld K, Franceschini N, Minke B. Evidence for a sensitizing pigment in fly photoreceptors. Nature. 1977;**269**:386-390

[49] Minke B, Kirschfeld K. The contribution of a sensitizing pigment to the photosensitivity spectra of fly rhodopsin and metarhodopsin. The Journal of General Physiology. 1979;**73**: 517-540

[50] Hardie RC. The photoreceptor array of the dipteran retina. Trends in Neurosciences. 1986; **9**:419-423

[51] Hamdorf K, Hochstrate P, Höglund G, Moser M, Sperber S, Schlecht P. Ultra-violet sensitizing pigment in blowfly photoreceptors Rl-6; probable nature and binding sites. J. comp. Physiol. A. 1992;**171**:601-615

[52] Hamdorf K, Schwemer J, Gogala M. Insect visual pigment sensitive to ultraviolet light. Nature. 1971;**231**:458-459

[53] Kashiyama K, Seki T, Numata H, Goto SG. Molecular characterization of visual pigments in Branchiopoda and the evolution of opsins in Arthropoda. Mol Biol Evol. 2009;**2**: 299-311. DOI: 10.1093/molbev/msn251

[54] Helmut L, Schmeinck S, Antonerxleben F. Identification and localization of visual pigments in the retina of the moth, *Antheraea polyphemus* (Insecta, Saturniidae). Cell and Tissue Research. 1986;**245**:81-89

[55] Tearle R. Tissue specific effects of ommochrome pathway mutations in Drosophila melanogaster. Genetical Research. 1991;**57**:257-266. PubMed PMID: 1909678

[56] Ozaki K, Nagatani H, Ozaki M, Tokunaga F. Maturation of major Drosophila rhodopsin, ninaE, requires chromophore 3-hydroxyretinal. Neuron. 1993;**10**(6):1113-1119

[57] JRM C, Casas M. The multiple disguises of spiders: A: Web color and decorations, body color and movement. Philosophical Transactions of the Royal Society B: Biological Sciences. 2009;**364**:471-480. DOI: 10.1098/rstb.2008.0212. PMC 2674075.PMID 18990 672

[58] Schwemer J. Visual pigments of compound eyes—Structure, photochemistry, and regeneration. In: Stavenga DG, Hardie RC, editors. Facets of Vision. Berlin, Heidelberg: Springer; 1989; pp. 134-151

[59] Insausti TC, Le Gall M, Lazzari CR. Oxidative stress, photo damage and the role of screening pigments in insect eyes. Journal of Experimental Biology. 2013;**216**:3200-3207. DOI: 10.1242/jeb.082818. PMID: 23661779

[60] Oxford GS, Gillespie RG. Evolution and ecology of spider colouration. Annual Review of Entomology. 1998;**43**:619-643. DOI: 10.1146/annurev.ento.43.1.619. PMID 15012400

[61] Takahiro O, Junya H, Keiro U, Ryo F, Toshiki T, Teruyuki N, Hideki S. A visible dominant marker for insect transgenesis. Nature Communications. 2015;**3**:1-9. DOI: 10.1038/ncomms 2312

[62] Goldsmith TH, Collins JS, Licht S. The cone oil droplets of avian retinas. Vision Research. 1984;**24**:1661-1671

[63] Bernard GD. Red-absorbing visual pigment of butterflies. Science. 1979;**203**:1125-1127

[64] Cromartie RIT. Insect pigments. Annual Review of Entomology. 1959;**4**:59-76. DOI: 10.1146/annurev.en.04.010159.000423

[65] Land M. The optical structures of animal eyes. Current Biology 2005;**15**:319-323. DOI: 10.1016/j.cub.2005.04.041

Chapter 2

Insect Conservation for the Twenty-First Century

Michael J. Samways

Additional information is available at the end of the chapter

http://dx.doi.org/10.5772/intechopen.73864

Abstract

Insects have been immensely successful as an animal group. They dominate compositional diversity of all but the saltiest and coldest parts of the planet. Yet today insects are declining at a precipitous rate. This is of great concern in terms of impoverishment of Earth, and is also dire for us. Insects contribute to the maintenance of terrestrial and freshwater systems, their service delivery and their resilience. The meteoric impact of humans is challenging this dominance, yet so few people realize that the very fabric of life on which they depend is being unraveled at an alarming rate. Action is required, as are new perspectives, if we are to maintain insect diversity and services through the twenty-first century. Here, we review how we should view and act to have more effective insect diversity conservation based on six themes: (1) philosophy (establishing the ethical foundation), (2) research (the finding out), (3) policy (the framework for action), (4) psychology (understanding how to engage humans in insect conservation action), (5) practice (implementation of action), and (6) validation (establishing how well we are doing at conserving insects). We then overview some emergent challenges and solutions at both the species and landscape operational levels in agricultural, forestry, and urban environments.

Keywords: insect conservation biology, insect species conservation, insect diversity, insect services, conservation strategies, caring for insects

1. Introduction

We live in a rapidly changing world. Yet the very fabric of life on which we depend is in jeopardy [1]. A major component of this fabric is the insects. Although they are small and rarely seen performing their myriad activities, they are critically important for maintaining the world as we know it [2]. They perform so many tasks that life without them would be a catastrophe. Yet, so few people even begin to realize just how important insects are in our everyday lives.

The aim here is first to overview insect success as the dominant organisms on the planet. Then, we focus on the threats that insects are currently facing as a result of human activity. Yet, time is short for us to do something about the escalating insect losses across the planet [3, 4].

While strategies are already in place for undertaking insect conservation, some are emerging as being crucially important for successful insect conservation into the twenty-first century, and beyond. We do this here by overviewing some emergent themes on which to base strategies for averting further insect losses. This is important, as insects are fundamental to terrestrial and freshwater ecosystem processes, and we need to maintain insect diversity for future human generations to appreciate, respect, and rely on for supplying essential services.

2. Insect success from a conservation perspective

Insects are the most speciose organisms on earth, making up 70% of all organisms. They dominate all but the coldest and saltiest environments. They inhabit deserts to tropical forests, and swampy pools to pounding waterfalls. They are the majority that few of us see, hidden in plain view. All terrestrial and freshwater plants, even mosses and liverworts, have associations with insects. Most plants have flowers, and their reproduction depends on insect visitors to pollinate them, and so reproduce. Virtually, all frogs and lizards need insects to sustain them. Well over half of all fish, birds, and small mammals require insect food. In turn, a third of insects eat other insects. In short, insects are the fundamental woof and weft of all land-associated ecosystems. Furthermore, we cannot live without them, as a third of our food, and especially the most nutritious components of our food, such as fruit and nuts, depends largely or totally on insect pollination.

Insect success has come about largely through the insect's body plan, with its three tagma (head, thorax, and abdomen), its immensely versatile skeletal structure, and highly varying physiology. The head is packed with a huge array of sensory apparatus. The thorax has highly effective legs for many environments, and, most notably, wings for dispersing and rapidly finding resources and mates. In turn, the abdomen, houses diverse digestive tracts, as well as reproductive apparatus that in some species produces millions of eggs. Indeed, reproductive potential can be extraordinarily high. Richard Harington calculated that when a gravid aphid is left to reproduce with no mortality, after 1 year, the earth would be covered 14.7 km deep in aphids!

Insects have a wide range of mouthparts molded from the robust chitin of the skeleton, so that they can chew, rasp, suck, and burrow through all sorts of organic tissue, living and dead, plant, animal, and fungal. They are also able produce an immense array of chemicals for attack, defense, camouflage, mate attraction, and digestion. Furthermore, their sensitivity to certain chemicals can be extraordinary, with some moths being able to detect a mate many kilometers away, and others, such as certain parasitoids, able to detect prey deep in plant tissue.

Insects are not just items, but also interactors. They are among the most ecologically connected of all organisms. A simple biotope with just 1000 species, leads to half a million potential interactions. This means that the over 1 million described insect species and the likelihood that there are about 5 million species in all, suggests that insects interact with virtually every component in the terrestrial and freshwater realms (**Figure 1**).

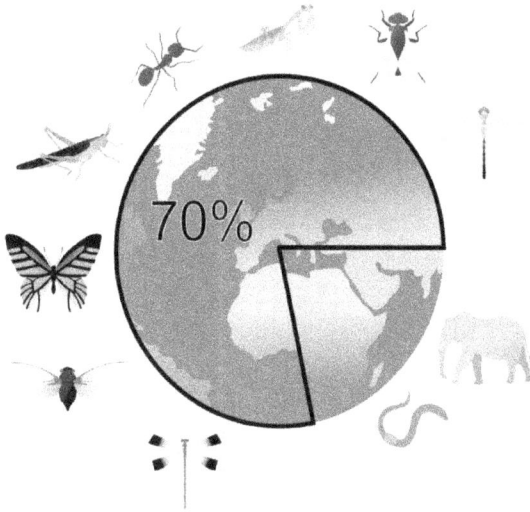

Figure 1. 70% of the species on Earth are insects, despite the land covering only a third of the planet. Insects in terrestrial and freshwater systems are the most highly ecologically connected of all organisms.

3. Insects in today's world

The world is in trouble, with the "World Scientists' Warning to Humanity: A Second Notice" [5] having been issued. We must act now, and decisively, on how we manage the planet. The Anthropocene ("age of humans") is well-established as the new geological era, and the sixth mass extinction is upon us [6]. Insects are central to how we react to this crisis, and how we should respond, as avoiding general ecocide in the twenty-first century rests on involving insects in the new world view.

What is of great concern is that insects appeared to have pulled through the last great extinction at the end of the Cretaceous, 66 million years ago, largely intact [7]. They also survived the various glacial maxima and minima by moving around to re-establish in thermal optima, and to some extent, independent of plants [8]. Such large-scale movement is not so feasible today. The human-induced patchwork of anthropogenic, novel ecosystems has created a myriad of barriers to free movement. With global climate change and landscape fragmentation being a "deadly anthropogenic cocktail" [9], the future for insect diversity depends on three options: (1) adapt on site, (2) move across the human-instigated barriers, or (3) die out. As (3) is not an ethical or survival option, either for insects or us, we must find ways that enable insects to survive through the twenty-first century and beyond.

Despite the importance of insects, it is only relatively recently that they have been mainstreamed into biodiversity conservation. This is being done at various operational levels from the species level through to the landscape level of conservation, with major decisions being made at the scale of nation states through National Biodiversity and Strategy Action Plans. Furthermore, as there are now global insect conservation initiatives, it highlights the adage

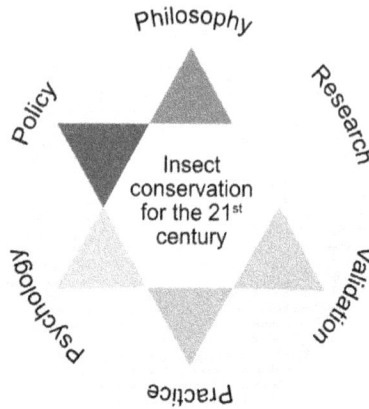

Figure 2. One way to view insect conservation into the twenty-first century is to focus on six, inter-related themes: philosophy, research, policy, psychology, practice, and validation.

"think global, act local." These operational levels, species, landscape, national, and global, are not mutually exclusive, but complementary.

Insect conservation in the twenty-first century can be seen against six inter-related themes: (1) philosophy (establishing the ethical foundation), (2) research (the finding out), (3) policy (the framework for action), (4) psychology (understanding how humans engage in insect conservation action), (5) practice (implementation of action), and (6) validation (establishing how well we are doing at conserving insects). We will now interrogate these themes in more detail. We do this against a background of species, landscape, national, and global operational levels, so as to move quickly to save the current insect diversity on Earth (Figure 2).

4. Philosophy for insect conservation

The starting point for insect conservation is to question why we should do it. Arguably, as extinction is the norm, with 99% of all organisms on earth having gone extinct from natural causes, perhaps we should just let events take their course, as a part of an evolving planet? There are two aspects here that we must consider to counter this view. First, there is the intrinsic value of insects, and that they must be conserved for their own sake, especially as they celebrate the immense complexity of life. The argument for intrinsic value is that we are sentient, and hopefully, as we have given ourselves the epithet *sapiens*, a wise and caring species. Quite simply, we share this lonely planet with an amazing variety of life and a stunning selection of insect forms. Are we so crass that we simply send them to oblivion? Second, and quite bluntly and selfishly, they have instrumental value, that is, they have value purely for us. Yet in reality, so few people actually appreciate this value.

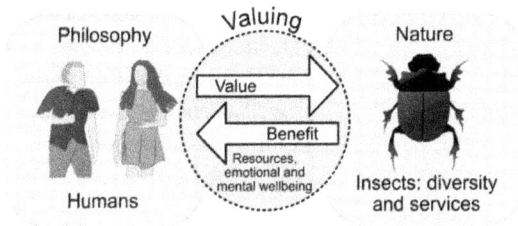

Figure 3. Justifying insect conservation, and then doing it requires a philosophical view based on valuing nature. When value is placed on the human relationship with nature, benefit accrues in terms of physical resources such as food, as well in terms of well-being, both emotional and mental.

With humanity having received its "Second Warning" and global ecosystems in major decline, we cannot carry on as we have been up to now. We require a radical change in thought and action. We need a good philosophical base for steering practice. Many people consider that conservation is instrumental and must benefit humans. This approach stems partly from the logic that by taking this approach, those who hold power will listen. In short, it is considered by many as the only hard currency of insect conservation.

The binary approach of intrinsic versus instrumental value of itself has shortcomings, as it focuses on entities, such as insect species or landscapes, rather than how we relate to nature. There is now a move away from this binary approach to one that focuses on personal and collective wellbeing, based on how we value and relate to nature to achieve this wellbeing [10]. This focus on relational value is built into our need for nature, and that we have a shared destiny, with biodiversity as a whole. This also means relating to insects as most of them are fundamental to our health and happiness, because without them we would have an impoverished and dangerous world as resources decline. Quite simply, we need to look after insects, and they will look after us. We can no longer ignore this fact if there is any future for our grandchildren (**Figure 3**).

5. Research needed for twenty-first century insect conservation

5.1. Operational levels of insect conservation research

Research is concerned with discovery of new information. For insect conservation, this research is about finding new and effective ways for maintaining insect diversity, insect species, and insect populations. As insects are embedded in the ecological fabric around them, and we need to understand it if we are to provide realistic insect conservation solutions, we research the optimal environmental conditions that enable insect survival. These environmental conditions may be abiotic, such as temperature regimes, fire frequencies and intensity, rainfall patterns and intensity, insolation, elevation, rockiness, water, pH, dissolved oxygen, as well as contaminants, pollutants, pesticides, and many others. Environmental variables are also biotic, including vegetation structure and composition, pollen and seed availability, fungal presence vs. absence, host availability (vertebrate or invertebrate, including other insects), mutualist presence, dung availability, mimic models, and so on.

As many insects express developmental polymorphism, where for example, the larva is morphologically and functionally very different from the adult, they may require a host of abiotic and biotic conditions and resources for optimal survival within their habitat. The larva of a butterfly needs a particular host plant(s), as well as enemy-free and disease-free space, while the adult needs certain nectar sources, mate-meeting sites such as hilltops, oviposition sites, sunny conditions for flight, besides enemy- and disease-free space.

Although an insect's habitat is embedded in an ecosystem, some require more than one ecosystem to sustain them [11]. With anthropogenic modification of the landscape, not only are conditions changed within their habitat, but also around it. The landscape matrix around a habitat may not only lack critical conditions and resources, so prohibiting generational survival there, but it also has an effect, often adversely, on the habitat itself. In terms of research, as well as investigating the habitat per se, we also must establish the landscape context and contrast. How we make this matrix more hospitable for insects is a major research thrust in this the twenty-first century. This is where we need to reconcile the needs of insects and those of humans. Progress is being made, but now we must hasten that process in the coming years.

Insects have to move and mate, and so maintain genetic diversity within a species, especially to maintain adaptability to changing conditions induced by humans. Functional connectivity across the landscape that facilitates movement, and so genetic exchange, has become a major challenge for the twenty-first century. Already much progress has been made. This has been done by making the human production landscape and urban environment more insect friendly. This has been done by putting in place stepping stone habitats, while perhaps not being optimal for long-term survival nevertheless provide stop over stations for insects as they move across the landscape, and by instigating conservation corridors, which as well as being for movement, are also source habitats that provide optimal conditions for all the life stages and the production of viable offspring.

5.2. Species-level insect conservation

From a conservation perspective, we can view "a species" as populations made up of a group of individuals. We do not actually conserve "a species" but individuals of the species. Groups of individuals in an overall population do, or do not, exchange occasional alleles. Those that do are metapopulations, and those that do not are subpopulations. Importantly, it is the range of genetic variation in populations and how it is shared among individuals that determines the adaptability of a population to environmental change, whether for the better or the worse, and whether in response to local (e.g., landscape fragmentation, pollution) or global (e.g., climate change) impacts, or both.

The viability of metapopulations depends on the flow of genes that provide high value for adapting to prevailing conditions. These conditions are currently changing rapidly, and are often adversely synergistic, with, for example, fragmentation and insecticidal impact together providing an even greater challenge over the impact of just one. Without gene flow, metapopulation dynamics can be disrupted, leading to an adapt-or-die situation. There might not be enough genetic variation within the isolated group of individuals for survival in the long term, because of changing or stochastic environmental events and/or genetic impoverishment. When populations are isolated by the effect of landscape fragmentation, there may some selection pressure to the new conditions, leading to human-induced evolution, known as anthropovicariance [12]. Philosophically, this leads to an interesting debate. Here, we are not conserving an existing natural phenomenon so much as creating a new one: to what extent is this "new insect" of conservation concern? This sort of philosophical challenge is what we now need to address in the twenty-first century. As novel landscapes (i.e., those created by humans) are now present, arguably we now need to conserve, or at least let live, those species with high adaptive ability, and therefore have an evolutionary future in this rapidly changing world [13].

Subpopulations present a different situation in that they already show some differentiation caused by natural drivers. These different subpopulations are known as evolutionarily significant units (ESUs), each unit of which deserves conservation in its own right. They are in effect evolution in action, and represent new species in the formation. However, some ESUs are threatened and others not. The English Large copper butterfly *Lycaena dispar dispar* is extinct, the Dutch ESU *L. dispar batavus* is highly threatened [14], and the Estonian *L. dispar rutilis* is common and expanding its geographical range [15]. The "species" has been re-introduced and has established in England from Europe, but this is not the original ESU. These are genetic and ethical issues that will confront us ever more this century.

The twenty-first century is likely to see much more focus on the genetics of species, bearing in mind this will always be about a few species attracting special attention. There will be several approaches, and these are already developing. Genetic work has shown that some species are very ancient, with the yellow presba dragonfly *Syncordulia gracilis* having a pedigree going back almost 60 million years [16]. Such species must receive conservation action if we are to show some empathy for ancient insects. Even resurrecting extinct species (revenant species) is feasible from well-preserved specimens in museums [17]. However, this is a lot of hard work for a privileged few in comparison with saving species by good and protective management of natural ecosystems in the first place (**Figure 4**).

5.3. Landscape-level insect conservation

Species conservation is arguably a luxury overlay on insect diversity conservation using landscapes, at least from a global perspective. Insect species conservation is morally right, but exclusive, and time is short for conserving as many insect species as soon as possible. However, there is a good reason to do good insect conservation in parts of the world that can afford that luxury, i.e., in those countries with high GDPs, and where there is great interest and involvement by the public. It leads to exploratory techniques and methods that we will, in the future, need globally. Meanwhile, for those countries with lower GDPs, as well as the more economically developed nations, conserving good quality landscapes with high habitat heterogeneity will not only conserve many insect species and their interactions all at the same time, but also will conserve many insect services to which the public and policy makers can both literally buy into.

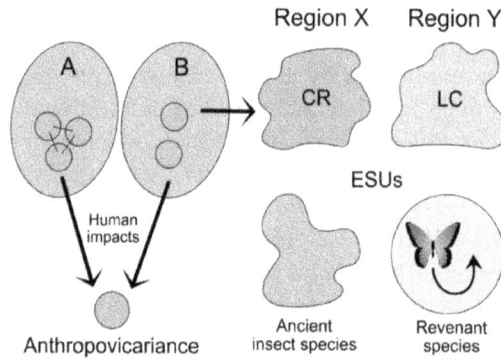

Figure 4. There are great opportunities for insect species-level conservation, as long as we have a good understanding of what "a species" stands for, based on genetic, behavioral, and ecological knowledge. Here (upper left) are two species (A and B), one (A) with populations that experience genetic exchange (metapopulations), and one (B) that does not (having two subpopulations). Either of these might undergo genetic change in response to human impacts, a phenomenon known as anthropovicariance (bottom left). Two subpopulations may be genetically, and usually morphologically, different, making up two evolutionarily significant units (ESUs) in two geographically distinct regions (X and Y) (upper right). The one on the left is red listed as critically endangered (CR), and of great conservation concern, while the one on the right is red listed as least concern (LC), which means that it is not of immediate concern, but it could always be threatened in the future. Ancient species, with a long phylogenetic pedigree, are often of high conservation significance as they are usually genetically highly irreplaceable (bottom middle). A revenant species is one that is extinct as a species and has been brought back to life from good quality genetic material extracted from museum specimens (bottom right).

Conceptually, we as humans relate to landscapes very well, as they fit comfortably into our frame of reference about nature. We see this already with the evolution of new perspectives. While there has been much progress in the past with species conservation, there is now a shift towards viewing nature as a vast array of benefits that it provides, which includes being in nature for our wellbeing.

Natural England's Conservation Strategy for the twenty-first century [18] articulates this well, and uses three guiding principles: (1) creating resilient landscapes (and seas), (2) putting people at the heart of the environment, and (3) growing natural capital (i.e., giving populations the chance to survive and increase). The earlier strategy of ring-fencing and protecting individual species and habitats has not been successful, having led to local species loss. The focus now is at the larger spatial and conceptual scales, with the development of resilient landscapes and ecosystems. This has led to research on the drivers of species loss and deterioration of ecosystems vis-à-vis what maintains species, their interactions, and ecosystem function. This means understanding what are opportunities and realistic strategies can be brought into play. These include engaging wildlife-friendly farming, gracing the urban environment with bio-diversity-friendly green spaces, and improving functional connectivity across the landscape.

All these biological perspectives must include people and their wellbeing if it is to succeed. The focus moves away from risk toward a new approach that involves enhancing and investing in the environment, leading to long-term stewardship of environmental assets. Interestingly, this new approach does not exclude a species focus, but rather integrates it into a vision of long-term resilience across landscapes. Crucially here, it includes insects of all types, whether

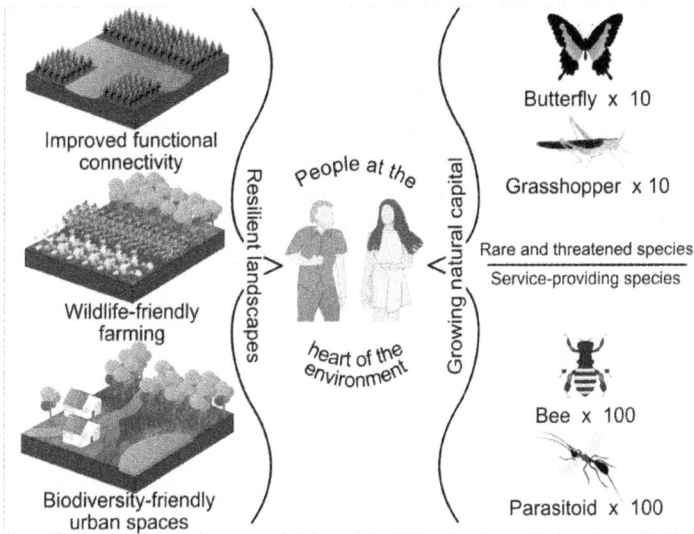

Figure 5. Landscape-level insect conservation based on three guiding principles: (1) creating resilient landscapes, (2) putting people at the heart of the environment, and (3) growing natural capital, as perceived by Natural England. Resilient landscapes include, for example, improved functional connectivity (here, grassland conservation corridors between exotic pine plantation tree blocks, wildlife-friendly farming such as using organic methods and leaving an increased proportion of natural habitat, as well as biodiversity-friendly spaces in the urban environment). Growing natural capital is just not only about improving indigenous species richness, but also improving the abundance of the focal species. In the case of rare and threatened species, an increase of ten times would be a great improvement, while for service-providing species, like bees for pollination and parasitoids for pest control, a 100-fold increase would be truly excellent.

threatened or not, which contribute to a more informed landscape approach. It is essentially a positive feedback loop, with all species, including humans, benefitting. In turn, it is the sensitive species on those landscapes that can tell us how well we are doing. These principles now require more research to tailor them to local circumstances (**Figure 5**).

6. Policy for insect conservation

There are two ways to consider insect conservation as regards policy. Insects are a component of biodiversity, and secondly, they provide essential ecosystem services. These perspectives can be considered at various spatial scales, from global down to national (bearing in mind that nation states are an important conservation action unit), and then local.

From a biodiversity perspective, insect conservation is integral biodiversity conservation, as insects function at various trophic levels, and so interact with many other organisms, plant, fungal, animal, protoctist, as well as among themselves. Biodiversity conservation is globally framed in terms of the Convention on Biological Diversity (CBD), which is agreed to by most nation states. Importantly, the Aichi Biodiversity Targets, under the umbrella of the CBD, provide some specific goals, and with the exception of those specifically relating to the marine environment, apply to insect conservation into the future. There are, however, a few marine insects, notably sea striders (Hemiptera, Gerridae: *Halobates* spp.).

Insects globally are conserved both at the species level and as insect diversity. At the species level, it is the International Union for the Conservation of Nature (IUCN)/Species Survival Commission (SSC) that is the major global proponent through production of the Red List of Threatened Species (www.iucnredlist.org). This globally recognized list provides just not only an inventory of the world's species and their threat status, but also gives suggestions for conservation action. It does this through the activities of a network of specialists on various taxonomic groups.

The red list however, is not a priority list for action, and nor is it a political document stipulating what governments or agencies should do to conserve insects. Quite simply, but importantly, it is an assessment of the conservation status of the insects that have so far been assessed. Nevertheless, it is highly influential for conserving insects globally. One reason for its impact is that once a species is assessed on the red list, and especially when a species in listed as threatened (as opposed to not, i.e., that it is classified as Least Concern), a species tends to become highly iconic. This is a great boost for listed insects, as they then receive the same treatment as a wombat, whale, or weasel. In short, their profile is raised considerably, and so automatically find their way into policy documents on biodiversity conservation.

The greatest challenge, among many [19], for insect species Red Listing is that the group is so speciose, with today only about 7700 species having been evaluated for the Red List, which is 1% of described species, 1,060,704 in all [20], and probably less than 0.2% of the millions that exist. The reason for these low percentages is that considerable field work is required to assess the threat status of an insect species, and there are relatively few insect specialists to do the job. The situation is aggravated by many species going extinct without even having received scientific names, a phenomenon known as Centinelan extinction (named after Centinela Ridge in Ecuador where botanists found many plants had gone extinct from deforestation before they could name them).

Of those insects that have been assessed, it is possible to get a sense of the types of threats facing them. When lumped with terrestrial invertebrates in general, the main threats in decreasing order are: (1) habitat loss due to logging, (2) habitat loss due to agriculture, (3) infrastructure development such as urbanization, (4) habitat loss and fragmentation due to transportation/service corridors, (5) invasive alien species, (6) change in fire regime, (7) pollution, (8) climate change/severe weather, and (9) mining [21]. For freshwater invertebrates, pollution and dams/poor management are the major threats, and greater than all the others together [22]. Knowing what these threats are not only helps us plan for the future, but also bearing in mind that for any one insect species, there may be more than one threat, as threats are often adversely synergistic.

International trade in insect species is regulated. This is done through the Convention on International Trade in Endangered Species (CITES). To date, however, few insect species are listed, and they are mostly large, highly collectable and charismatic species, like birdwing butterflies. Objectively, there should be many more insect species that are CITES listed, but this is not the case because the limited financial resources for doing so are prioritized for restricting trade on charismatic vertebrates and rare plants, which are facing a desperate plight in its own right.

Moving down to the national scale, insect species conservation varies greatly from one country to another, with those countries with a high Gross National Product usually devoting more attention and resources to insect species conservation than the financially more constrained countries. Some countries have strong regulation protecting insects (e.g., all dragonflies in Germany). There are also consortia, and especially notable is the European Union Habitat Directive, where there is co-operation on protecting insects across the continent.

As regards, ecosystem services as defined by the Millennium Ecosystem Assessment [23], insects are highly significant. In terms of provisioning services, they are important as biological control agents of pests, for monitoring ecosystems, as well as for providing new medicines, and acting as tourist attractions in the wild, and in commercial butterfly houses in urban areas. Regulating services provided by insects include nutrient cycling, pollination, seed dispersal, stopping, or slowing invasions by other insects, and contribution to atmospheric gases. They also provide supporting services through breaking down living and dead plant material, as well as turning over soil. In turn, cultural services are many, and include representativeness of the variety of life, connecting with the natural world (especially through children), and use in genetic research.

All these services are often strongly recognized by policy makers, and regularly form the basis of justification for conserving insects, i.e., for what they do for us, which in the USA is estimated to be $57 billion/year [24]. But, we must be careful not to confuse the conservation of insects for the services that they provide with the conservation of insect species, which may not be the same set [25].

With recognition of the value and importance of insects with each passing year, insects are being increasingly enshrined in policy, at least in countries that have great respect for nature and recognize that our survival depends on valuing and conserving them. Recognition of the value and importance of insects will become more important with the passing of time this century.

Insects have been eaten by humans (entomophagy) from the earliest of times, with today an estimated 2 million people eating insects as part of their traditional diet. This is because insects are highly nutritious, being high in protein, essential fatty acids, and in important minerals such as calcium, copper, iron, selenium, and zinc. Commonly eaten insects are beetles, moth larvae, bees, wasps and ants, cicadas and other bugs, termites, dragonflies, flies, as well as some other insects [26]. Entomophagy will become increasingly important in the twenty-first century, both for direct human consumption and for livestock. Entomophagy is also now being used as a way of tackling malnutrition in children [27].

This increased reliance on insects is partly driven by red meat production being three times more resource hungry than insect production. While insects traditionally have been harvested from the wild, there is now a move to rear them on a large scale. This intense farming of insects has challenges, but new rearing and processing techniques and methods will inevitably come about simply because natural habitats are decreasing, and the size and demands of the human population are increasing. Improved insect farming will go hand in hand with new developments in the visual appearance of the food, and its design, so as to make insect food increasingly acceptable to a more discerning human population [28] (**Figure 6**).

Convention on Biological Diversity

Aichi Biodiversity Targets

Species
Global: Red List
National: National Red Lists
NBSAP's

Services
MEA
(Provisioning, Regulating, Cultural)
Entomophagy

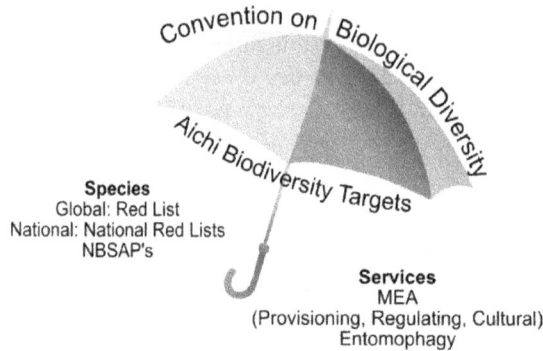

Figure 6. Insect conservation policy has various perspectives. Overarching is the Convention on Biological Diversity to which most countries in the world are signatories. The Aichi Biodiversity Targets, which are specific targets, mostly have great application to insect conservation. In terms of conservation of species *per se*, the International Union for the Conservation of Nature's global red list gives important coverage to insects. At the national level, local red lists (RLs) are important too. National Biodiversity Strategy Action Plans (NBSAPs) are also significant for local insect conservation, especially in biodiversity hotspots. NBSAPs cover a wide range of biodiversity issues and targets besides insect species conservation. As regards services supplied by insects, these may be described under the Millennium Ecosystem Assessment (MEA). It provides a framework to include provisioning, regulating and cultural services. An important service currently being supplied by insects is entomophagy, the human consumption of insects, which is becoming increasingly mainstream globally.

7. Insect conservation psychology

Understanding human behavior, promoting wellbeing, and increasing human care for nature is conservation psychology [29], which recognizes our dependence on nature and an understanding of why and how it is so essential for our wellbeing. Given, the huge role that insects play in providing services of all sorts, insect conservation psychology is going to play an important, if not vital, part in our survival into the twenty-first century [30].

As we advance technologically, we seem oblivious of the collateral damage we are doing to so many other species and their interactions, despite so many warnings that we are vitally dependent on them. In short, we are doing untold damage to the very systems on which we depend, with insects being at the heart of them, at least on land and in freshwater. We simply take nature, and insect services, for granted, and this cannot continue into the twenty-first century without dire consequences. Just as we do not see water vapor coming off plants, or the oxygen being breathed out by them, we just do not see insect diversity and our dependency on it: out of sight, out of mind.

Many of us have great interest and even respect for insects as children. But there are only certain "species," like "the" butterfly, bee, grasshopper, ladybird, and dragonfly that we hold dear. These are the iconic insects that are fascinating to us as children (and adult entomologists!), and are also benign. Only as we grow up, do we begin to realize that "the fly" is dirty, and "the mosquito" is a nuisance, if not dangerous. So, all insects get lumped together in our adult perceptions, i.e., not worthy of our consideration and conservation. This realization is now changing for the better, with the global realization that we have a pollinator crisis around the world, and "the bees are dying" [31]. This has aroused considerable awareness of our dependency on insects.

Insect conservation psychology

Insects

Conservation

Wellbeing

Humans

Biophilia
(learned)
Protected areas, parks,
botanical gardens, home
gardens

Citizen science

Insect icons

Children — Adults

Fear factor
a. Biophobia
(intrinsic)
b. Loss of
essential
insect services

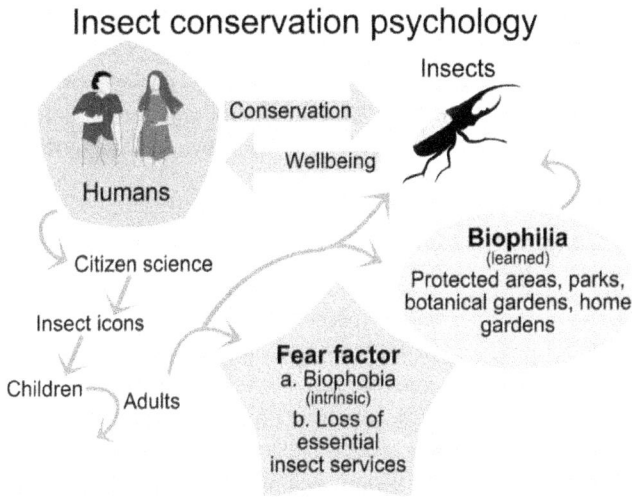

Figure 7. Insect conservation psychology is the relationship between humans and insects. The conservation of insects is essential for our wellbeing. The involvement of the wider body of the public in recording, monitoring, and engaging in conservation is citizen science. This activity promotes wellbeing in people while helping ensure a future for insects. Insect icons like "the bee, butterfly, and ladybird" are highly significant for children, but often we as adults forget them in our preoccupied lives. This is partly because an awareness develops that "the fly and the mosquito" are not good for us. We are also aware that the wasp stings, and so will the bee if we are unkind to it. This is biophobia (a), which is intrinsic to us. Yet when we overcome this fear factor, we culturally develop biophilia, especially when we see the beauty of nature, including insects, in our parks and gardens. Besides biophobia, the fear factor today has a second component, a concern that there is a loss of essential services (b), particularly pollination. Our reaction to this service loss, alongside biophilia for certain species, is feeding back into insect conservation, which, in turn, improves our sense of wellbeing.

It has also dawned on us that we need to be far more aware that widespread use of insecticides is not only detrimental to our health but also undermining our very food base, despite actually having been articulated in the 1960s [32]. The pollination crisis has also led to much action to find solutions to it, and for once, right now without delay [33]. Like climate change, the pollinator crisis is no longer something to think about in the late twenty-first century—it has already arrived with full force.

While our dislike of many insects is intrinsic (biophobia/entomophobia), love and appreciation of insects (biophilia/entomophilia) is learned [34]. This learning can only come about from positive experiences, for example, by seeing insects in their natural environment, whether in the home garden, botanical garden, or protected area. Indeed, protected areas receive 8 billion people visits per year [35], with a profit of $250 billion/year [36], making protected areas, along with other natural areas, good for human wellbeing, while also earning their keep and supporting a vast array of insects.

Linking insect conservation psychology to research and policy can come about through citizen science, the involvement of informed and enthusiastic sectors of the public for recording the distribution of species and engaging in insect conservation [37]. This involvement of the public has greatly improved our knowledge of many insect groups [38–40]. The improvement in knowledge, and hence insight, comes about simply because there are more enthusiastic eyes and hands in the field increasing the amount of information on species, and this can be very high in the case of threatened, red listed species, as well as many others that are not threatened, at least for now. This information then also becomes valuable for lobbying policy makers. Enthusiastic citizen scientists are also making direct contact with policy makers, and therefore increasing the voice of change-for-better in insect conservation. This approach is going to become increasingly important in the twenty-first century (**Figure 7**).

8. Insect conservation in practice

Insect conservation is a practical activity based on good research and a sound knowledge base. The foundation for practice is a well-understood philosophical underpinning, valuable research findings, and then a strategic response to well-informed policy, alongside applied insect conservation psychology. Yet, the practice of insect conservation is rarely actually carried out by philosophers, researchers, policy makers, or psychologists. Insect conservation action is usually carried out by conservation organizations, concerned farmers, and progressive private companies, and other responsible land stewards and owners, all of whom are highly cognizant of philosophy, research, policy, and psychology. It is also carried out by citizen scientists, many of whom may be associated with self-grown non-governmental organizations like the Xerces Society, Butterfly Conservation, and Buglife, who receive donations from private funders, as does IUCN at the global scale. Citizen scientists are usually active in their home area, where they carry out insect conservation activities in addition to monitoring, especially on nominated insect species, and often those of special concern. This is another reflection of the "think global, act local" principle.

There is a baseline for insect conservation that is non-negotiable, that is, the recognition of formally proclaimed protected areas composed of natural habitat. The significance of this is that these areas are often an island in a mosaic of novel landscapes. They are frequently the last bastion for many rare, specialist and/or highly localized endemics, as well as for specialized interactions. As protected areas are usually isolated, they may require some management, usually to mimic the historic condition. Introduction of fire regimes, for example, may be required to simulate the situation before modern human fragmentation of the landscape. Management of fire may be to reduce the fuel load, or so reduce risk of an intense and potentially highly damaging fire, or it may be to stop vegetation succession to an unnaturally woody environment [41].

There are approaches that also intergrade protected areas with the surrounding novel mosaic. Among these are biosphere reserves, of which there 669 in 120 countries, and ratified by the United Nations Educational, Scientific, and Cultural Organization. Biosphere reserves consist of a core and two surrounding zones. The core of the biosphere reserve is fully protected and managed according to historic conditions. Surrounding that is a buffer zone with only low-level human impact on the biota and ecosystem function. Outside the buffer zone is the transition zone, which supports sustainable agriculture or forestry based on agro-ecological principles or organic agriculture, as well as having little infrastructure development.

The transition zone is characterized by low insecticide use, maximal use of biological control, and cultural control measures of pests, low compaction practices (i.e., avoiding heavy machinery),

Figure 8. (Left) An organic vineyard with no insecticide input, mulching of the inter-rows, and the planting of the inter-rows to biodiversity-friendly vegetation. These vineyards are particularly rich in soil fauna as well as in above-ground insect diversity. (Right) Seen here is a large-scale ecological network of conservation corridors of remnant, natural, high value grassland in and among plantation forestry using alien pine trees. These grassland corridors not only conserve biodiversity but also maintain hydrological processes in a natural state.

planting of indigenous flora, and the setting aside of remnant patches as reserves per se or as stepping stones to protected areas. Despite the great importance and high level of instigation of biosphere reserves, there is still little research on their effectiveness for conserving insects and their interactions. Research and validation of these reserves is going to be increasingly important in the twenty-first century, as they may well prove to be a major initiative for harmonizing human activity for optimal production and effective insect conservation (**Figure 8**).

Other insect-friendly approaches are also being used. These go hand in hand with agro-ecological approaches [42]. They follow a spectrum from land sharing (the mixing of crops and insect-friendly plants, e.g., coffee plantations in natural forest [43] to the separation at a larger spatial scale between conservation areas and production areas, known as land sparing (e.g., instigation of interconnected conservation corridors of remnant land to form large-scale ecological networks, e.g., grassland and natural forest patches among plantation forestry blocks [44]. These approaches have been very successful, with ecological networks effectively extending the size of the adjacent protected area and providing much more resilience to the system overall [45].

Originating in Europe, but now more widespread, has been the development of agri-environment schemes, which at the start involved financial compensation to farmers for setting aside land and not cultivate it [46]. These schemes have now grown into fully agro-ecological areas which aim to maximize opportunities for indigenous biodiversity while optimizing production. The aim is not only to conserve insects and other wildlife, but also to promote natural ecosystem services such as pollination and biological control of pests [47, 48].

In general, embracing these new approaches has been beneficial for insect conservation, despite farmers often being concerned that production might dip when moving across to agro-ecological/organic approaches from conventional practices [49]. Nevertheless, the future is looking bright for these new eco-friendly approaches, with increasing global pressure to use farming and forestry techniques that conserve insects and promote human health. These approaches are going to become increasingly prevalent this century. They will be especially effective when high-intensity agriculture is converted into a softer approach, which may require some restoration of both terrestrial and freshwater ecosystems.

Already, more than half the human population lives in cities, with the proportion increasing as the century progresses. There are two fundamental issues to consider: human intensification of the urban landscape, and increasing disconnection between humans and nature. Intensification leads to proportionately less space for nature, and increasing pressures on what remains. Only the widespread generalist insect species will likely survive. As habitat space becomes scarcer, all insects will also be less abundant. Furthermore, there are going to be some major genetic change among those that do not die out.

In this new urban space, there must be provision for all the life stages. While butterfly gardens usually provide nectar for adults, they must, in addition, provide food plants for the larvae. Metapopulation dynamics must also be maintained, and while this may be possible for some species in large urban parks, there will need to be conservation corridors, often known as greenways in an urban setting, to maintain genetic diversity within each species.

Pressures on insect populations in the urban environment are also great [50]. Temperature increase in the urban environment, the heat island effect, presents one challenge. Increased road kill from vehicles and above-ground transportation is another. Pollution is an issue for humans and insects alike. For insects, another major impact is the huge effect of lights, especially on night flying insects.

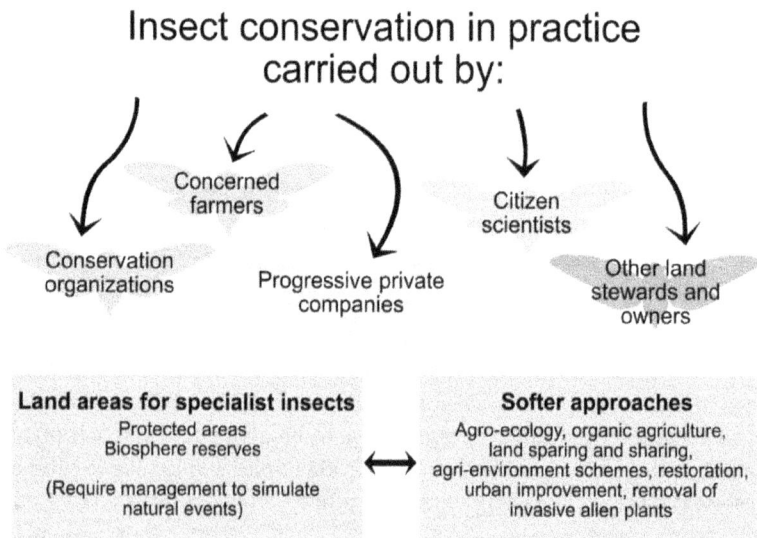

Figure 9. Insect conservation in practice is carried out by various organizational communities. Land cores, such as protected areas and the core zones of biosphere reserves, are critical for many specialist insect species. These areas do, however, sometimes require some management to simulate natural events. Outside of these land cores, softer approaches to landscape design and management take place as insect conservation action. These approaches include agro-ecology, organic agriculture, the land sparing-land sharing spectrum, agri-environment schemes, restoration, urban improvement, and removal of invasive alien plants.

There are areas of mitigation that are likely to become increasingly important in the twenty-first century. As regards the two fundamental issues, needs of humans as well as of insects, both are inter-twined. Good quality habitat space, terrestrial and freshwater, must be available, and largely populated with indigenous vegetation, and without invasive alien plants. Green walls and roofs are also likely to increase greatly, but mostly as carbon sinks and for esthetics than for native insect diversity. Nevertheless, this would increase awareness and contribute to rescuing the extinction of experience [51], where people, especially children, can stay connected with nature. However, with the increasing devastating impacts of global change leading to more extreme weather events, safety and security must be integrated with an awareness of the plight of insects. However, care for insects is likely to slip down the list of priorities as extreme weather events become more prevalent (**Figure 9**).

9. Validation

Our fifth theme is validation, which links back to the first, research [44]. We need to check how well we have done after we have actually engaged in insect conservation. We measure our insect conservation success by using a range of variables that may be abiotic (e.g., water flow and soil chemistry) or biotic (e.g., plant diversity, plant health, insect diversity, and recovery of insect populations). Before the conservation action, we would have determined the goal. What exactly are we aiming to conserve? Is it a particular species? If so, will it pull through adverse weather conditions? Will it be able to tolerate or move with climate change? Or, we

Figure 10. Validation for improved insect conservation means that we have understood the fundamental conservation ethics, and have recognized the relevance of various policies. We then set the conservation goal(s) and undertake the research. We further engage insect conservation psychology as we put research findings into practice. Then comes a critical point in the circle: validation. Here, we ask how well we are doing in implementing research, and whether the conservation practice is effective. If not then we undertake more research. This validation/research/practice cycle is actually never ending, as we must monitor even if we think what we have put into practice is working, as conditions may change, especially given climate change and always the risk of a new stochastic event.

may aim to conserve insect diversity using conservation corridors among plantation forestry blocks. Is the insect diversity the same as measured against a benchmark such as in a nearby protected area? Although species richness may be the same, is the composition the same? Or, when we engage agro-ecology, are the focal services, whether pollination with bees or pest control using natural enemies, being adequately supplied? If so, are the local rare and threatened species also being conserved?

Validation is a circular process: setting the conservation goal, establishing a time line, determining resources to achieve the goal, validating the goal, identifying shortfalls, understanding those shortfalls with research, putting in place a new goal which has then been refined, and so on. In short, as the twenty-first century progresses, we are going to focus more on a healthy and reliable environment, and this is going to require accountability in the form of validation (**Figure 10**).

10. Conclusions

Time is short for conserving the necessary insect diversity to sustain us. Future generations will decry this era of despoliation of the planet, and what we have done to the web of life, including the vast beauty, grace, intricacy, and worthiness of insects. Insects and other biodiversity symbolize this blue and white jewel in an almost incomprehensible vastness of molecular simplicity that we call space. Once the novelty of landing on Mars has worn off, humans will look back at Earth and realize with a great awakening how special it is. This may well (hopefully) trigger a renewed enthusiasm and effort to conserve the great poetry of biodiversity, especially that of insects. It will be a true case of "the grass was not, after all, greener on the other side of the fence."

We are improving our ethical base, but too slowly. We have rapidly increased insect conservation research, but too slowly. We have put in place policy, but acting on it too slowly. We are developing insect conservation psychology, but too slowly. And, as for action, it is being done pitifully too slowly. We are beginning to instigate validation processes, but still too slowly.

We will achieve much insect conservation this century, surfing on the wave of becoming scared of what a terrible mess we are making of the planet and now that our very life base is threatened. We are developing improved range of technologies, from conservation genetics through to satellite technology and information flow, all of which will help us make progress. But we still have to do the physical hard work on the ground to get it all right. The more we live in a virtual world, the more disconnected from nature we will be. We will wake up one day and it will dawn on us that the fiction of money [52] will not sustain us. Nature will always survive, with Conservation International's message "we need nature, nature does not need us" becoming totally real. We will kill many insects, but those which do survive will exist for a lot longer than we will. What saddens me personally is that our grandchildren will say "why did grandpa and grandma destroy so much without thinking of us?"

Acknowledgements

Special thanks to Charl Deacon for preparing the figures. Funding was from Mondi Group.

Conflict of interest

The author declares no conflict of interest.

Author details

Michael J. Samways

Address all correspondence to: samways@sun.ac.za

Department of Conservation Ecology and Entomology, Stellenbosch University,
South Africa

References

[1] Dirzo R, Young HS, Galetti M, Ceballos G, Isaac NJB, Collen B. Defaunation in the Anth-ropocene. Science. 2014;**345**:401-406

[2] Samways MJ. Insect Conservation Biology. London: Chapman and Hall; 1994. 358p

[3] Mawdsley NA, Stork NE. Species extinctions in insects: Ecological and biogeographical considerations. In: Harrington R, Stork NE, editors. Insects in a Changing Environment. London: Academic Press; 1995. pp. 321-369

[4] Hallmann CA, Sorg M, Jongejans E, Siepel H, Hofland N, et al. More than 75 percent decline over 27 years in total flying insect biomass in protected areas. PLoS One. October 18, 2017:1-5. DOI: 10.137/journal.pone.0185809

[5] Ripple WJ, Wolf C, Newsome TM, Galetti M, Alamgir M, Crist E, Mahmoud MI, Laurance WF, et al. World scientists' warning to humanity: A second notice. BioScience. 2017;**67**:1026-1028. DOI: 10.1093/biosci/bix125/4605229

[6] Ceballos G, Ehrlich PR, Barnosky AD, Garcia A, Pringle RM, Palmer TM. Accelerated modern human-induced species losses: Entering the sixth mass extinction. Science Advances. 2015. E1400253 June 2015:1-5

[7] Labandeira CC, Johnson KR, Wilf P. Impact of the terminal cretaceous event on plant-insect associations. Proceedings of the National Academy of Sciences USA. 2002;**99**:2061-2066

[8] Ponel P, Orgeas J, Samways MJ, Andrieu-Ponel V, de Beaulieu J-L, et al. 110 000 years of quaternary beetle diversity change. Biodiversity and Conservation. 2003;**12**:2077-2089

[9] Travis JMJ. Climate change and habitat destruction: A deadly anthropogenic cocktail. Proceedings of the Royal Society of London B. 2003;**270**:467-473

[10] Chan KM, Balvanera P, Benessaiah K, Chapman M, Dìaz S, et al. Why protect nature? Rethinking values and the environment. Proceedings of the National Academy of Sciences, USA. 2016;**113**:1462-1465

[11] Knight TM, McCoy MW, Chase JM, McCoy KA, Holt RD. Trophic cascades across eco-systems. Nature. 2005;**437**:880-883

[12] Williams BL. Conservation genetics, extinction, and taxonomic status: A case history of the regal fritillary. Conservation Biology. 2002;**16**:148-157

[13] Casacci LP, Barbero F, Balletto E. The "Evolutionarily Significant Unit" concept and its applicability in biological conservation. The Italian Journal of Zoology. 2014;**81**:182-193

[14] Barnett LK, Warren MS. Species Action Plan: Large Copper *Lycaena dispar*. Colchester, Essex, UK: Butterfly Conservation; 1995. p. 42

[15] Lindman L, Remm J, Saksing K, Sober V, Õunap E, Tammaru T. *Lycaena dispar* on its northern distribution limit: An expansive generalist. Insect Conservation and Diversity. 2015;**8**:3-16

[16] Ware JL, Simaika JP, Samways MJ. Biogeography and divergence time estimates of the relic South African Cape dragonfly genus Syncordulia: Global significance and implications for conservation. Zootaxa. 2009;**2216**:22-36

[17] Prosser SW, Dewaard JR, Miller SE, Hebert PDN. DNA barcodes from century-old type specimens using next-generation sequencing. Molecular Ecology Resources. 2016;**16**:487-497

[18] Conserva 21. Natural England's Conservation Strategy for the 21st Century. 2016. 11p. www.gov.uk/natural-england

[19] Cardoso P, Erwin TL, Borges PAV, New TR. The seven impediments in invertebrate conservation and how to overcome them. Biological Conservation. 2011;**144**:2647-2655

[20] Foottit RG, Adler PH, editors. Insect Biodiversity: Science and Society. 2nd ed. Oxford, UK: Wiley-Blackwell

[21] Gerlach J, Hoffman BS, Hochkirch A, Jepsen S, Seddon M, et al. Terrestrial invertebrate life. In: Collen B, Böhm M, Kemp R, JEM B, editors. Spineless: Status and Trends of the World's Invertebrates. London, UK: Zoological Society of London; 2012. pp. 46-57

[22] Darwall W, Seddon M, Clausnitzer V, Cumberlidge N. Freshwater invertebrate life. In: Collen B, Böhm M, Kemp R, JEM B, editors. Spineless: Status and Trends of the World's Invertebrates. London, UK: Zoological Society of London; 2012. pp. 46-57

[23] Synthesis Team Co-chairs: Duraiappah AK, Naeem S. Millennium Ecosystem Assessment Ecosystems and Human Well-Being: Biodiversity Synthesis. Washington DC, USA: World Resources Institute; 2005. 25p

[24] Losey JE, Vaughan M. The economic value of ecological services provided by insects. BioScience. 2006;**56**:311-323

[25] Kleijn D, Winfree R, Bartomeus I, Carvalheiro LS, Henry M, et al. Delivery of crop pollination services is an insufficient argument for wild pollinator conservation. Nature Communications. 2015;**6**:7414

[26] van Huis A, Van Itterbeeck J, Klunder H, Mertens E, Halloran A, et al. Edible Insects: Future Prospects for Food and Feed Security. Food and Agriculture Organization of the United Nations: Rome, Italy; 2013. 187p

[27] Mmari M. Can feeding young children a porridge made from insects improve their health status? Ruforum. Newsletter. 2017;**1**:9-12

[28] Chung J, Aguirre-Bielschowsky J. Ento: Introducing edible insects into the Western diet. Antenna. 2014;**38**(1):10-15

[29] Clayton S, Myers G. Conservation Psychology. Understanding and Promoting Human Care for Nature. Wiley-Blackwell: Oxford, UK; 2009. 253p

[30] Simaika JP, Samways MJ. Insect conservation psychology. Journal of Insect Conservation. 2018 (in press)

[31] Potts SG, Biesmeijer JC, Kremen C, Neumann P, Schweiger O, Kunin WE. Global pollinator declines: Trends, impacts and drivers. Trends in Ecology & Evolution. 2010;**25**:345-353

[32] Carson R. Silent Spring. Greenwich, Connecticut, Boston, USA: Fawcett; 1962. 378p

[33] Klein S, Cabirol A, Devaud J-M, Barron AB, Lihoreau M. Why bees are so vulnerable to environmental stressors. Trends in Ecology & Evolution. 2017;**32**:268-278

[34] Simaika JP, Samways MJ. Biophilia as a universal ethic for conserving biodiversity. Conservation Biology. 2016;**24**:903-906

[35] Balmford A, Green JMH, Anderson M, Beresford J, Huang C, et al. Walk on the wild side: Estimating the global magnitude of visits to protected areas. PLoS Biology. 2015;**13**:1-6. DOI: 10.1371/journal.pbio.1002074

[36] McCarthy DP, Donald PF, Scharlemann JP, Buchanan GM, Balmford A, et al. Financial costs of meeting global biodiversity conservation targets: Current spending and unmet needs. Science. 2012;**338**:946-949

[37] Roy DB, Ploquin EF, Randle Z, Risely K, Botham MS, et al. Comparison of trends in butterfly populations between monitoring schemes. Journal of Insect Conservation. 2015;**19**:313-324

[38] Lewandowski EJ, Oberhauser KS. Butterfly citizen scientists in the United States increase their engagement in conservation. Biological Conservation. 2017;**208**:106-112

[39] Zapponi L, Cini A, Bardiani M, Hardersen S, Maura M, et al. Citizen science data as an efficient tool for mapping protected saproxylic beetles. Biological Conservation. 2017;**208**:139-145

[40] Domroese MC, Johnson EA. Why watch bees? Motivations of citizen science volunteers in the Great Pollinator Project. Biological Conservation. 2017;**208**:40-47

[41] New TR. Insects, Fire and Conservation. Springer, New York, USA: Springer; 2014. 208p

[42] Tscharntke T, Tylianakis JM, Rand TA, Didham RK, Fahrig L, et al. Landscape moderation of biodiversity patterns and processes—Eight hypotheses. Biological Reviews. 2012;**87**:661-685

[43] Perfecto I, Vandermeer J. Biodiversity conservation in tropical agroecosystems: A new conservation paradigm. Annals of the New York Academy of Sciences. 2008;**1134**:173-200

[44] Samways MJ, Pryke JS. Large-scale ecological networks do work in an ecologically complex biodiversity hotspot. Ambio. 2016;**45**:161-172

[45] Pryke JS, Samways MJ. Ecological networks act as extensions of protected areas for arthropod biodiversity conservation. Journal of Applied Ecology. 2012;**49**:591-600

[46] Tscharntke T, Klein M, Kruess A, Steffan-Dewenter I, Thies C. Landscape perspectives on agricultural intensification and biodiversity—Ecosystem service management. Ecology Letters. 2005;**8**:857-874

[47] Carvalheiro LG, Kunin WE, Keil P, Aguirre-Gutiérrez J, Ellis WN, et al. Species richness declines and biotic homogenisation have slowed down for NW-European pollinators and plants. Ecology Letters. 2013;**16**:870-878

[48] Donald PF, Evans AD. Habitat connectivity and matrix restoration: The wider implications of agri-environment schemes. Journal of Applied Ecology. 2006;**43**:209-218

[49] Schneider MK, Lüscher G, Jeanneret P, Arndorfer M, Ammari Y, et al. Gains to species diversity in organically farmed fields are not propagated at the farm level. Nature Communications. 2014;**5**:4151

[50] New TR. Insect Conservation and Urban Environments. New York, USA: Springer; 2015. 244 p

[51] Samways MJ. Rescuing the extinction of experience. Biodiversity and Conservation. 2007;**16**:1995-1997

[52] Harari YN. Homo Deus: A Brief History of Tomorrow. Penguin Random House: London, UK; 2017. 513p

Economic Importance and Role of Sex Peptide

Understanding Perturbation in Aquatic Insect Communities under Multiple Stressor Threat

Alexa C. Alexander

Additional information is available at the end of the chapter

http://dx.doi.org/10.5772/intechopen.74112

Abstract

In the scientific literature, there is a considerable consensus that working toward evaluating multiple stressors is worthwhile. Unfortunately, our means to evaluate the combined effects of multiple stressors on species is limited. In agricultural systems, the relative threat posed to aquatic insect communities due to individual stressors (e.g., individual insecticides) is relatively well understood. However, understanding mixtures of pesticides, let alone the addition of complex and potentially interacting, natural gradients (e.g., nutrients and predation), is far harder. The objective of the following review was to evaluate the individual and combined effects of a range of multiple agricultural stressors on aquatic insect communities using a series of seven outdoor mesocosm experiments conducted since 2003. The mesocosm studies show that macroinvertebrate community responses can be similar, subtle, or even opposing depending on the stressors investigated and the mechanistic or ecological focus of the study. The current focus on individual chemicals and responses to treatment is misleading. Cumulative effects and multiple sublethal stressors are the norm in impacted ecosystems. A simple, holistic approach to environmental risk assessment is needed.

Keywords: aquatic communities, multiple stressors, mesocosm experiment, multiple predator theory, insecticides, nutrients, benthic macroinvertebrates, insect predators, review, synthesis

1. Introduction

Streams draining agricultural watersheds contain complex mixtures of pesticides, nutrients, and sediment due to runoff, spray drift, and erosion [1]. Pesticides also tend to be present at sublethal concentration levels at which we even know less about the cumulative toxicity and

IntechOpen

multiple stressor threat of mixtures of substances [2]. Some estimates suggest that >50% of river miles in the continental United States include mixtures of five or more pesticides, moderate to highly enriched nutrients and sediments [3]. More recent work has reported similar trends reporting the widespread use of insecticides and neonicotinoids in particular [4–6].

The exposure to mixtures of insecticides and other compounds pose a particular risk to aquatic insects because target biochemical receptors in insects are highly conserved [7]. For instance, the nicotinic acetylcholine receptor (nAChR), the primary binding site for neonicotinoid insecticides in insect pests, has been reported in numerous insect orders (e.g., Hemiptera, Blattodea, Homoptera, Orthoptera, and Diptera) [7]. Among the most highly publicized nontarget species affected by neonicotinoid insecticides are bees (*Apis mellifera*) [8]. Similarly, aquatic insects, such as mayflies (Order: Ephemeroptera), are also negatively affected by exposure to neonicotinoid insecticides at levels associated with agricultural runoff [9, 10]. Responses in other orders of aquatic insects, such as insect predators (e.g., Plecoptera and Odonata), are less studied but preliminary data suggest that these compounds likely affect a wide range of taxa. Knowledge gaps in our understanding of keystone taxa such as predators may have serious implications for risk assessment as density, and trait-mediated responses may have cascading effects on other members of aquatic food webs [11].

In the literature, there is a considerable consensus that working toward evaluating multiple stressors is worthwhile and important [12–14]. However, there has been virtually no uptake in addressing multiple stressors in ecological risk assessment. This may be due to the complex results emanating from mixture studies, which can be challenging to interpret [15]. Mixture studies are also typically retrospective and rarely address likely combinations of substances [16]. More proactive approaches that examine intentional or unintentional overlap in the field application of chemicals are needed.

The objective of the following studies was to evaluate the effect of multiple, interacting, natural, and anthropogenic stressors on aquatic macroinvertebrate communities. Responses primarily focus on the effects of the neonicotinoid insecticide imidacloprid, individually and in combination, with environmentally relevant mixtures of other substances and changing ecological conditions. Seven mesocosm studies were conducted between 2003 and 2010. Tests included exposure (individually and in mixture) to the following compounds: imidacloprid, the fungicide chlorothalonil, and the organophosphorus insecticides chlorpyrifos and dimethoate. Natural gradients were also examined and included changes in nutrient gradients such as low, medium, and high nutrient enrichments (oligotrophic, mesotrophic, and eutrophic) and increased predation pressure (added stonefly and dragonfly nymphs). Unique to this work is the comparison between responses of aquatic communities tested over time to overlapping treatments all collected from the same riverine source (see Materials & Methods). Further, concentrations selected were within the range of concentrations of pesticides and nutrients that have been detected in runoff and offer new insights as to why some streams become degraded. These findings have never before been summarized; thus, collectively, the following represents a unique snapshot of the range of effects of multiple agricultural stressors on aquatic insect communities.

2. Materials and methods

2.1. Study species

Benthic insects live on the bottom of streams and interact with multiple environmental compartments including water, sediment, and gravel interfaces [17]. Benthic macroinvertebrates (BMI) are good indicators of stream health because changes in BMI diversity and abundance can be associated with some contaminants [18]. Aquatic insects, like midges (Order: Diptera) and mayflies (Order: Ephemeroptera), lend themselves to studies of nutrients and contaminants since they both share many life history characteristics and yet are sufficiently different to highlight changes in streams. Midges in our streams were dominated by the family Chironomidae. Chironomids are small-bodied (adults: 1.5–20 mm [19]) with a short life cycle and emerge throughout the spring, summer, and fall in Atlantic Canada (unpub. data). Like many mayflies, chironomids are often members of the collector-gatherer or scraper trophic guilds, feeding on benthic algae, bacteria, and organic matter. Mayflies are larger than chironomids and may take prolonged periods to develop with some mayfly families only able to emerge once a year [20]. Mayflies are also generally considered to be sensitive to stress, in contrast to the more tolerant midges, and can be good indicators of contamination.

Aquatic insect predators such as dragonflies and stoneflies have also been shown to be sensitive to changes in habitat condition and agricultural gradients, particularly, nutrients [21]. As aquatic nymphs, dragonflies and stoneflies are highly opportunistic predators and show strong allometry to the average body size of their prey [22]. *Gomphus borealis* (Odonata and Gomphidae) are ambush predators that burrow in sediment to await the arrival of suitable prey items [23]. These generalized predators [24] feed by ejecting their labium to grasp their prey before devouring them. In contrast, *Agnetina capitata* (Plecoptera and Perlidae) are foraging predators [25] and search mechanically for prey.

2.2. Study site and allocation of treatments

Since 2003, mesocosm experiments have been conducted at the Environment and Climate Change Canada mesocosm test facility located at Agriculture and Agri-Food Canada, 10-km southeast of Fredericton (New Brunswick, Canada). Among these experiments were a series of studies conducted to examine the effects of multiple stressors on aquatic macroinvertebrate communities. These studies were designed to test the additive, cumulative, and interactive effects of the insecticide imidacloprid, in mixtures of similar (e.g., three insecticides) and dissimilar (insecticide and fungicide) chemicals on aquatic insect assemblages. Test conditions manipulated concentrations of insecticides (imidacloprid, dimethoate, and chlorpyrifos), fungicides (chlorothalonil), nutrients (oligo-, meso-, and eutrophic gradients) and predation pressure (stoneflies and dragonflies). In brief, the chemicals tested were chlorpyrifos (O,O-Diethyl O-(3,5,6-trichloro-2-pyridinyl) phosphorothioate) and dimethoate (O,O-Dimethyl S-[2-(methylamino)-2-oxoethyl] phosphorodithioate) both organophosphorus insecticides that are among the top 10 most commonly used in North America as well as being highly toxic to nontarget aquatic species [26, 27]. Imidacloprid (1-((6-Chloro-3-pyridinyl)methyl)-N-nitro-2-imidazolidinimine) is a neonicotinoid insecticide, while chlorothalonil (2,4,5,6-tetrachloro-1,3-benzenedicarbonitrile) is a widely used fungicide in Atlantic Canada [28, 29].

The experiments were designed to evaluate a range of conditions (**Table 1**) for example, (1) a chronic, low nutrient (oligotrophic) study conducted in the Fall of 2003 (22 September 2003–21 October 2003) that explored continuous exposure to the insecticide imidacloprid in the lethal effects range; (2) a pulse, low nutrient (oligo-mesotrophic boundary) study conducted in the Summer of 2004 (20 June 2004–10 July 2004), which combined a chronic and a pulse experiment that explored lower concentrations of the same range of insecticide exposures with the addition of some nutrients (e.g., [TN] 25 ± 3 µg/L) described in [10]; (3) a pulse, meso-trophic nutrient enrichment study conducted in the Fall of 2004 (3 August 2004-1 September 2004) that included the addition of moderate nutrients (as above and [TN] 30 ± 4 µg/L); (4) a pulse, low nutrient study conducted in the Fall of 2005 (4 August 2005–24 August 2005) and an imidacloprid-chlorothalonil mixture experiment that explored the same range of insec-ticide exposures and nutrients see [30]; (5) a binary (1:1) mixture of two insecticides chlor-pyrifos and dimethoate (12 July–2 August 2007) [31]; (6) a ternary (1:1:1) mixture of three insecticides chlorpyrifos, dimethoate, and imidacloprid (16 August–6 September 2009) [21]; and (7) a pulsed imidacloprid within a nutrient gradient study conducted in 2010 (17 July–6 August 2010) see [32].

For each study, 80 artificial streams or outdoor mesocosms (**Figure 1**) were inoculated with a benthic macroinvertebrate community collected in the Nashwaak River, New Brunswick,

Experiment	Exposure duration in -d or -h	Stressors tested (ppb)	References
1. Chronic (press), oligotrophic study	20-d	Imidacloprid (5, 15)	
2. Press vs. pulse, oligo-mesotrophic study	20-d or 12-h	Imidacloprid press (0.1, 0.5, 1) and pulse (0.1, 0.5, 1, 5, 10)	[10]
3. Sublethal (pulse), mesotrophic study	24-h (2×) or 24-h (4×)	Imidacloprid (0.5, 1)	
4. Pesticide mixture (pulse), oligo-mesotrophic study	24-h (3×)	Imidacloprid (0.6, 17.6) Chlorothalonil (3, 30)	[30]
5. Insecticide mixture (pulse), oligo-mesotrophic study	96-h (1×)	Chlorpyrifos (1, 2, 4) Dimethoate (5, 10, 20)	[31]
6. Insecticide mixture (pulse), oligo-, and mesotrophic study	96-h (1×)	Imidacloprid (0.5, 1, 2) Chlorpyrifos (0.5. 1, 2) Dimethoate (2, 4, 8)	[21]
7. Nutrient-insecticide (pulse), oligo-, meso- and eutrophic study	96-h (1×)	Imidacloprid (1.4, 5)	[32]

All experiments were conducted over a 20-d period. Concentrations of stressors tested given in parts per billion (ppb), throughout.

Table 1. Overview of the design of seven mesocosm experiments conducted between 2003 and 2010.

Canada (46°8'34.584" N × 66°22'1.992" W). Each flow-through stream was circular and had a planar area of 0.065 m² and a 10-L volume. Each treatment level contained at least eight replicate streams. Treatment levels varied depending on the test objective but are summarized in detail elsewhere (see **Table 1**). Throughout, chemical analyses determined the actual concentrations of pesticides (National Laboratory for Environmental Testing, ECCC Saskatoon) and nutrients (RPC Fredericton). In brief, pesticide analyses were conducted on a Micromass Quattro Ultima liquid chromatography mass spectrometer (LC-MS/MS) with Waters 2695 Alliance HPLC System equipped with a Waters Xterra MS C18 (100 × 2.1 mm i.d., 3.5 µm particle size, Milford, MA, USA) analytical column. Samples were routinely collected on multiple occasions during and after the exposure period. Pesticide samples were stored in 500 ml amber vials (EPA vials, Fisher scientific, Fair Lawn, NJ, USA) and stored at 4°C until shipment to Saskatoon for analysis. Nutrient treatments were chosen based on Biggs [33] and corroborated using in-stream chlorophyll-*a* measurements compared to levels reported in Dodds et al. [34]. Water quality samples and emergent insects were collected daily throughout each experiment.

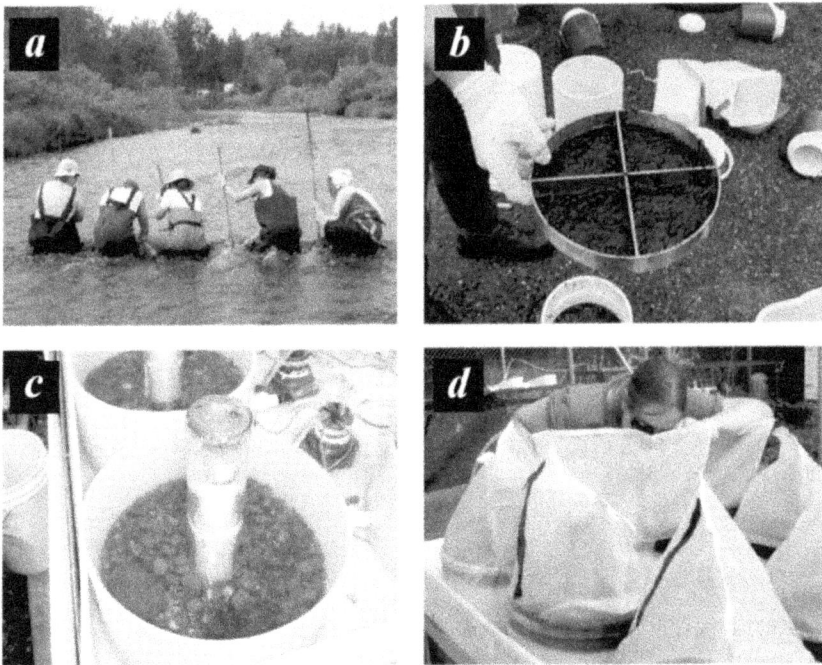

Figure 1. Outdoor, flow-through, stream mesocosms. (a) Benthic macroinvertebrates are collected by five samplers collecting 4 U-nets each. (b) The benthic community is then subsampled (four-way pie-plate subsampler shown). Community subsamples are then inoculated into replicate streams (e.g., ¼ of community sampled per replicate). (c) Each replicate stream is circular (0.065 m² and 10-L volume) and was also inoculated with five cobblestones and coarse and fine gravel. (d) After inoculation with benthic macroinvertebrates each stream is covered with 45 µm mesh to facilitate the daily collection of emergent insects.

2.3. Ecological endpoints

At the end of each 20-d mesocosm experiment, the streams were dismantled and the contents collected. Water samples, periphyton samples, and invertebrates were collected from each replicate stream. For chlorophyll-*a* (μg/cm^2) and ash-free dry mass (AFDM, mg/cm^2), three scrapings (each 60.2 cm^2) were collected into 20-mL scintillation vials and frozen in a portable freezer at −20°C (Engel fridge/freezer MT35F-U1, Sawafugi Electric Co. Ltd., Tokyo, Japan). Aquatic nymphs and emergent adults were then measured using the Auto-Montage imaging program (Syncroscopy, Synoptics Inc., Frederick, MD, USA) with a Leica digital camera and dissecting microscope (Leica Microsystems Ltd., Cambridge, UK). Multiple photographs were taken of each organism and measurements were conducted on segments using linear and curvilinear measurement tools. Calibrations were conducted for each objective lens and were repeated for individual insect measurements if the coarse or fine focus was adjusted. Numerous measurements were taken, including maximum head length and width, maximum thorax length and width, wing pad length, and total body length. In the absence of wing pads, the total length of the thorax was measured from the center of the anterior tip of the pronotum dorsally to furthest posterior point along the centerline of the metanotum. When wing pads were present, the total length of the thorax was measured from the center of the anterior tip of the pronotum dorsally to furthest posterior tip of the wing pad along the left lateral axis. Predation pressure was estimated as the product of the density (per cm^2) and body size (mm) of predators such as the stonefly *Agnetina capitata* and dragonfly *Gomphus borealis* per replicate stream (described in [32]).

2.4. Statistical analysis

Responses were examined using a complement of standard parametric (ANOVA) and multivariate statistical tools including: (e.g., nonmetric multidimensional scaling, factor analysis, principal components analysis as well as mixed general linear and structural equation models) see [35–37]. Assumptions of statistical tests were met throughout. Differences in river subsamples and control tanks were assessed using the Euclidean distance method to compare the distance of reference samples calculated by the unweighted pair group method [38]. Structural equation models (SEM) were used to assess changes in food webs between treatment levels and were estimated using covariance in partial regression coefficients [39]. Finally, principal components analysis was used to confirm the strength of relationships due to nutrient treatment.

3. Results

Seven mesocosm experiments were conducted between 2003 and 2010 (**Table 1**). Responses varied between studies but the pesticide or nutrient treatment applied were major drivers of changing patterns in the macroinvertebrate community. Changes over time due to successional or seasonal changes in the sampled aquatic community were less evident than those due to pesticide or nutrient treatment. For instance, at the onset of the mesocosm experiments, subsampled river communities were similar to other subsamples collected during the same period (**Figure 2**). River communities were also similar to assemblages observed in control streams at the end of the 20-d mesocosm experiment (**Figure 2a**). However, treatment with neonicotinoid insecticides such as imidacloprid (5 or 15 ppb, 20-d press exposure) resulted in

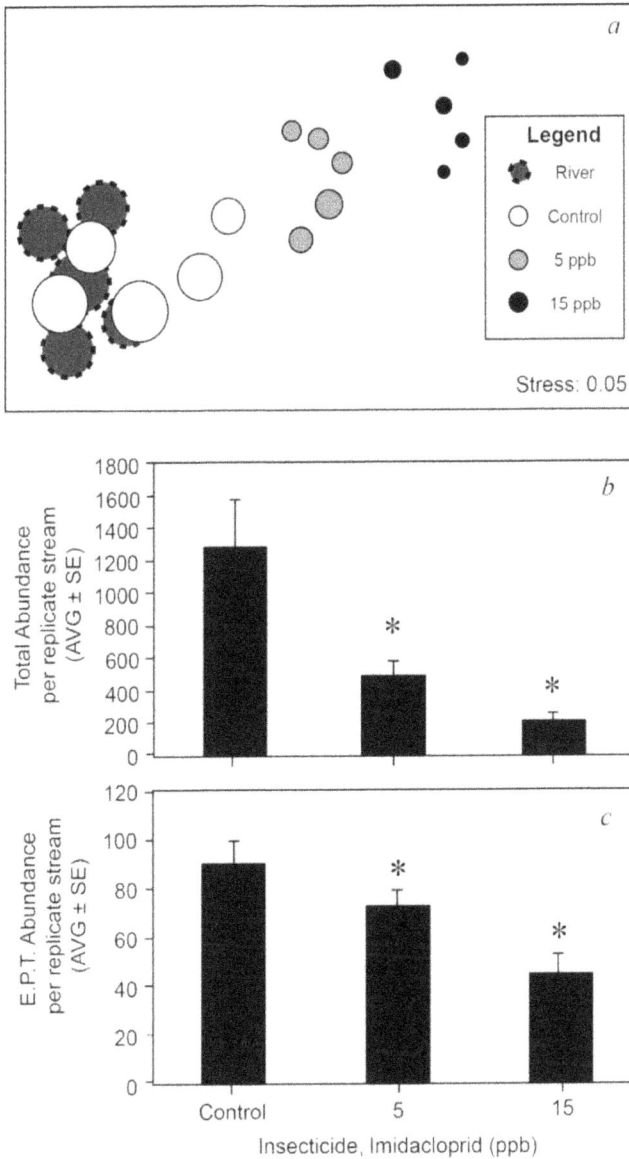

Figure 2. Benthic macroinvertebrate community responses (a) nonmetric multidimensional scaling (nMDS) to treatment with three concentrations (control, 5 and 15 ppb) of the neonicotinoid insecticide, imidacloprid, in the 2003 pilot mesocosm study. The size of the circles reflects the abundance of organisms and the distance between circles the magnitude of change between replicate communities. (b) Total abundance of aquatic macroinvertebrates per replicate stream (AVG total no. per stream ± SE). (c) Abundance of sensitive E.P.T. (orders Ephemeroptera, Plecoptera, and Trichoptera) aquatic insect taxa per replicate stream (AVG total of E.P.T. only per stream ± SE). Significant differences (P < 0.05) are indicated (*).

major changes in the abundance and diversity of aquatic insect taxa (**Figure 2a**). For example, severe reductions (>78 and 92% in 5 and 15 ppb) in the total abundance of taxa (**Figure 2b**) and sensitive E.P.T. taxa were strongly associated with imidacloprid treatment (>18 and 49%;

see **Figure 2c**) (e.g., $F_{3,30} \geq 5.43$, $P < 0.01$). Further, experiments examining an increasing range of imidacloprid concentrations demonstrated similar and significant decreases in community total abundance, total richness, and E.P.T. abundance (e.g., Mesocosm #1, $F_{2,14} \geq 5.90$, $P \leq 0.01$; Mesocosm #4: $F_{2,71} \geq 3.30$, $P \leq 0.05$) (**Table 1**).

Nutrient treatment also differed between studies (**Table 1**). Enrichment could be measured as changes in periphyton abundance (as chlorophyll-*a* in µg/cm²) and was consistent with the nutrient treatment applied (low to high enrichment: oligo-, meso-, or eutrophic). Responses to nutrient enrichment were consistent irrespective of the year of study or seasonal changes in the macroinvertebrate community. Community responses to the combined action of nutrients and insecticides could also appear similar. For instance, the removal of insect grazers (structural change) at the base of the food web in high insecticide treatments was associated with increased periphyton biomass (functional change). Thus, oligotrophic streams treated with imidacloprid were more similar to mesotrophic or even eutrophic conditions due to grazer release despite the lack of nutrient enrichment (e.g., 3.3 ± 0.5 µg/cm² due to 15 ppb treatment with imidacloprid) ($F_{2,23} = 3.91$; $P = 0.03$).

A factor analysis of benthic macroinvertebrate community responses to treatment explained 45% of the variance in all of the community data collected between 2003 and 2009 (Cumulative Eigenvalue 21.38) (**Figure 3**). Throughout, responses to treatment differed ($P < 0.05$) between Factor 1 (E.V. 17.68 of 21.38, 37%) and Factor 2 (E.V. 7.73 of 21.38, 45%). Factor 2 was closely correlated with the magnitude (concentration × duration) of imidacloprid concentration ($r = 0.65$, $P < 0.05$) and Factor 1 reflected differences associated with community composition (e.g., presence, absence, and diversity). In control streams, macroinvertebrate community responses to oligotrophic and mesotrophic enrichment overlapped, whereas responses to eutrophic treatment were discernibly separated from those in lower levels of nutrient enrichment (**Figure 3a**). Treatment with a single insecticide also overlapped for similar chemical compounds such as the insecticides imidacloprid, dimethoate, and chlorpyrifos ($P > 0.05$) (**Figure 3b**). In contrast, community responses to dissimilar chemicals, such as mixtures of imidacloprid and nutrients, diverged from those of imidacloprid alone (**Figure 3c**). Community responses also diverged in response to the combined action of imidacloprid, nutrients, and increased predation pressure (**Figure 3c**). Interestingly, community responses to mixtures of imidacloprid and the fungicide chlorothalonil were similar despite differences in the mode of action of these two compounds (**Figure 3c**).

A structural equation model of the covariant relationships between different organisms, trophic guilds, and other metrics (e.g., periphyton biomass) was also used to compare food webs in the nutrient enriched (mesotrophic) versus limited (oligotrophic) streams (**Figure 4**). In oligotrophic streams, only two response variables significantly covaried ($P < 0.05$) (**Figure 4a**). Specifically, the density (no./cm²) of the dragonfly *Gomphus borealis* covaried with ash-free dry mass, or AFDM (mg/cm²), but did not covary with the density of other predators, such as the stonefly *Agnetina capitata* or the abundance of scrapers (**Figure 4a**). Rather, the density of *A. capitata*, covaried with scrapers ($P < 0.05$), which in turn may be associated with chlorophyll-*a*, but only at the $P < 0.1$ level. In contrast, mesotrophic streams had 17 covariant relationships ($P < 0.05$) between different taxa and guilds (**Figure 4b**).

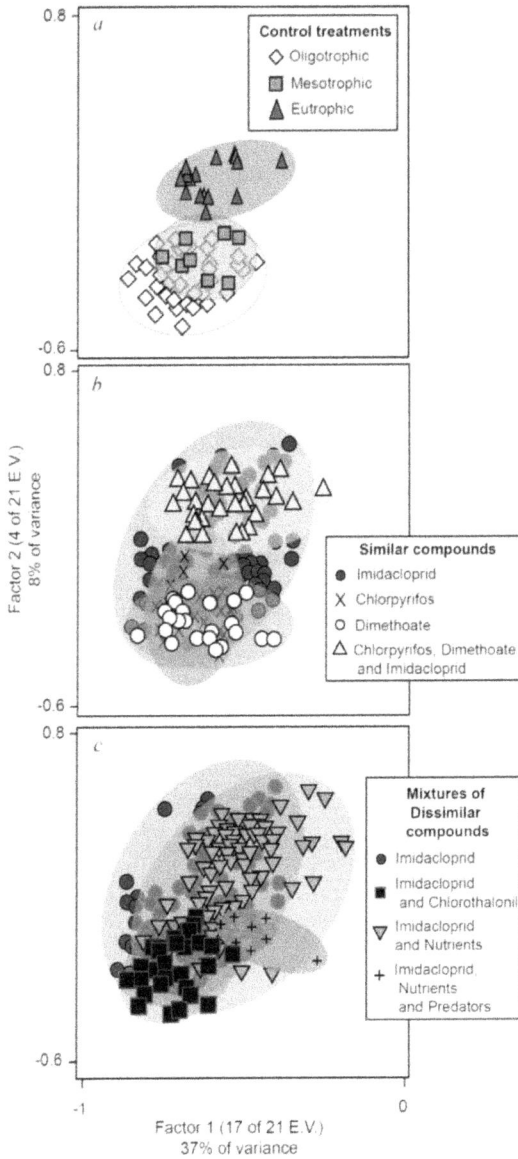

Figure 3. Factor analysis of benthic macroinvertebrate community abundance (no. of different genera per treatment level) during 7 years of mesocosm experiments subdivided into (a) control treatments with the addition of no nutrients (oligotrophic), moderate nutrients (mesotrophic), and high nutrients (eutrophic). (b) Exposure to similar insecticides either individually (imidacloprid, chlorpyrifos, and dimethoate) or in mixture (all three insecticides), and (c) exposure to mixtures of dissimilar chemical contaminants (as mixtures only). Dissimilar contaminants tested included the insecticide imidacloprid, and fungicide chlorothalonil, imidacloprid in the presence of nutrient enrichment (mesotrophic or eutrophic) and imidacloprid in the presence of mesotrophic nutrients and stonefly predators. Ellipses enclose all replicate treatment responses at the 95% CI. Lack of overlap between ellipses suggests statistically significant differences between responses to treatment at the $P < 0.05$ level.

For instance, the density of *G. borealis,* covaried (P < 0.05) with the density of its main competitor, *A. capitata,* as well as with other predators. Collectively, *G. borealis* and *A. capitata* both covaried with the density of a range of taxa including consumers from multiple sensitive orders (E.P.T. consumers), as well as scrapers, collector-gatherers, and shredders (**Figure 4b**). In turn, these taxa, and collector-gatherers in particular, affected the density of other guilds (e.g., E.P.T. consumers, collector-filterers, and piercers) as well as the standing stock of the periphyton community (AFDM, chlorophyll-*a*) (**Figure 4b**). Eutrophic conditions were only examined in a single mesocosm study (#7, conducted in 2010, see **Table 1**), and as such, relationships between taxa and guilds are less generalizable than those reported for oligotrophic and mesotrophic streams.

Responses, however, within eutrophic streams overlapped those in oligotrophic and mesotrophic nutrient treatments as well as with specific stressor conditions unique to Mesocosm #7, the only eutrophic gradient tested (**Figure 5** and **Table 1**). Genera and guilds

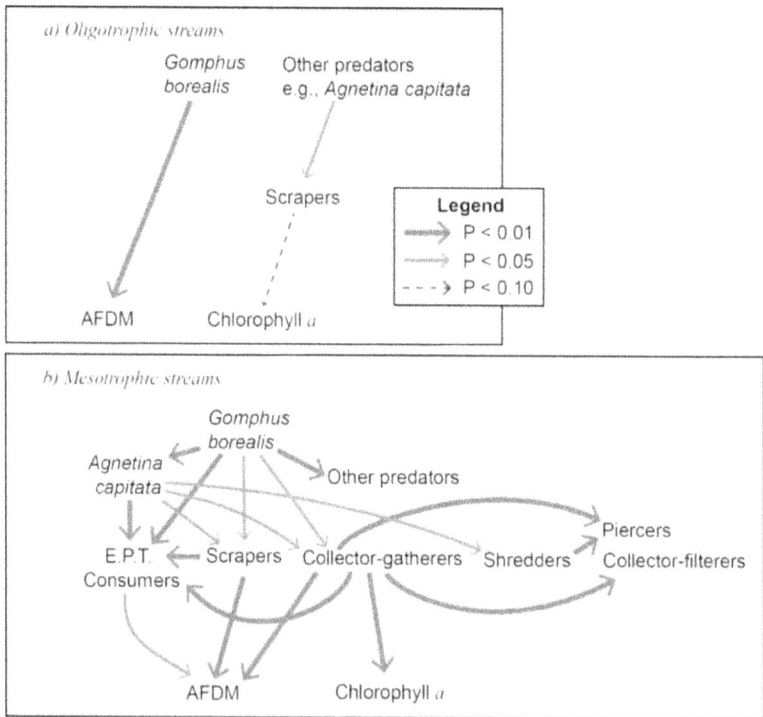

Figure 4. Summary of significant covariant relationships between the density (no./cm²) of different taxa, guilds and other metrics in control streams under oligotrophic (a) or mesotrophic (b) nutrient treatment. (a) Only two significant covariant relationships were reported under nutrient limited (oligotrophic) conditions whereas under (b) moderately nutrient enriched conditions (mesotrophic), 17 covariant relationships between taxa, guilds, or periphyton biomass were evident (measured as chlorophyll a in μg/cm² (chlorophyll) were found.

tended to respond similarly to treatment, and 68% of the variance in macroinvertebrate density could be explained by treatment with nutrients or the insecticide imidacloprid (52.3% of Factor 1 and 15.9% of Factor 2, **Figure 5**). For instance, total abundance, E.P.T. abundance, total richness, and density of collector-gatherers were all primarily (r ≥ 0.72, Factor 1) responding to the combined action of nutrient and insecticide gradients and secondarily to nutrient treatment specifically (r ≤ 0.63, Factor 2). In contrast, chlorophyll a and AFDM were only highly correlated (r = −0.68 and r = −0.71) to Factor 2. Finally, communities in control eutrophic streams were most similar to oligotrophic streams that were simultaneously treated with concentrations that are lethal to 50% of the insect population (median lethal concentration or LC50). Thus, in eutrophic streams, concentrations that would be highly significant stressors in less enriched streams were closely related to responses associated with baseline condition in these highly enriched systems (**Figure 5**).

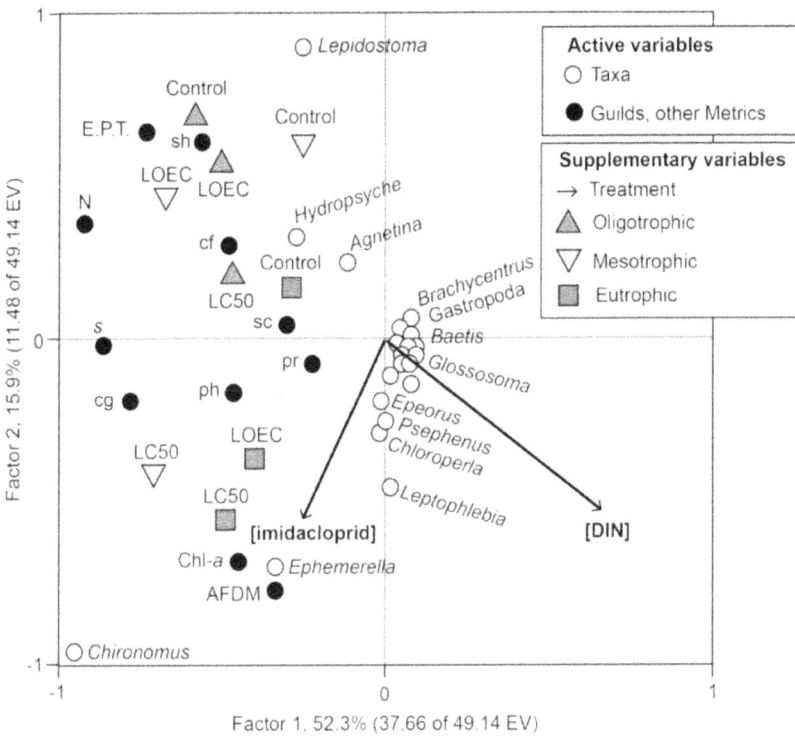

Figure 5. Principal components analysis of mesocosm 7 only (17 July to 6 August 2010) explaining 60% (52.3 + 15.9% EV) of the variation in benthic macroinvertebrate community (no./stream/cm²) and periphyton biomass (chl-*a* in μg/cm² and AFDM mg/cm²) due to either nutrient enrichment (oligotrophic, mesotrophic or eutrophic due to the addition of dissolved inorganic nitrogen [DIN]) or neonicotinoid insecticide treatment (imidacloprid as control, lowest observable effect concentration [LOEC], or median lethal concentration [LC50]). Density of select genera and guilds are highlighted; for example, total abundance (N), richness (s), E.P.T. abundance (E.P.T.), collector-filterers (cf), collector-gatherers (cg), piercers (ph), predators (pr), scrapers (sc) and shredders (sh). All comparisons were made using a correlation matrix.

4. Discussion

Streams draining agricultural catchments contain complex and often sublethal mixtures of pesticides and nutrients [1]. Ecological risk assessments rarely consider chemical mixtures, let alone combinations of natural and anthropogenic gradients. Regulators focus on individual compounds. Pesticides are regulated in Canada using a risk ranking approach based on an evaluation of the presence of available application data (e.g., sales or max application rate), chemical fate information (e.g., persistence and mobility), and toxicity (e.g., single species toxicity tests on fish, invertebrates, or aquatic plants). This focus on mortal responses to individual compounds poses a problem because it fails to consider conditions that are common in the environment: sublethal mixtures of chemicals are widespread. It is also evident that single species laboratory tests of individual compounds cannot approximate mixtures of chemicals affecting interacting assemblages of organisms in ecosystems.

The results of the studies described above show that in combination, pesticides and nutrients can reshape food webs (see also [9, 10, 21, 30–32]). In isolation, the action of these stressors appears to supersede underlying seasonal differences in macroinvertebrate communities. This finding suggests that nutrients and pesticides are fundamental drivers of effects in impacted aquatic communities. However, macroinvertebrate responses to pesticides and nutrients were varied and responses may be structurally similar yet functionally different. In the studies described above, responses due to nutrients and insecticides, such as the neonicotinoid and imidacloprid, were difficult to discern. The removal of grazers (**Figure 2**) at the base of the food web also increased periphyton biomass to levels that would suggest moderate or even high levels of enrichment (> 3 µg/cm^2) despite the lack of added nutrients (Mesocosm #1, in 2003). Further support for this finding is found in a separate experiment (Mesocosm #7, in 2010) where eutrophic streams were structurally and functionally similar to nutrient-limited streams simultaneously dosed with lethal doses (LC50) of imidacloprid (**Figure 5**). Collectively, these findings suggest that cascading effects at one end of the food web are common but could be due to different, and potentially, interacting pathways.

At lower doses, community responses to stress tended to overlap (**Figure 3b**) [10, 21]. For instance, communities were structurally similar due to low dose mixtures of three insecticides (chlorpyrifos, dimethoate, and imidacloprid) or due to any of these same compounds when tested individually at moderate or even high doses (**Figure 3b**). However, differences in community structure could be subtle as responses to treatment with mixtures of different types of compounds (e.g., pesticides vs. nutrients) tended to have less overlap when co-exposed to either substance individually (**Figure 3c**). Further evidence for structural changes in aquatic communities due to nutrients is apparent in the structural equation model (**Figure 4**). The covariant relationships between taxa varied widely between nutrient enriched versus limited streams despite the same aquatic macroinvertebrate assemblage being initially introduced into each treatment level.

Varied responses to different types of chemical compounds may appear to make ecological risk assessment difficult (see Kienzler et al. [16] for a review of approaches). Currently, in Canada, risk rankings list the toxicity of chemical compounds to different types of taxa (invertebrates, fish, or plants) using data collected from single species toxicity tests. For instance, fish toxicity ranks include different pesticides than rankings developed for invertebrates or plants. Specifically, the top three pesticides that are thought to pose the greatest risk to invertebrates are the neonicotinoid insecticides imidacloprid, thiamethoxam, and clothianidin. These same neonicotinoids are ranked as being far lesser risk of toxicity to fish (9, 20, and >30) or plants, respectively (>30). At present whether these substances are likely to co-occur is not considered.

There are advantages to the joint testing of substances. For instance, by testing effects jointly the number of tests to be conducted may decrease as only relevant mixtures need testing. Joint testing will also deepen our understanding of dose-dependent effects of similar and dissimilar mixtures of chemical compounds offering new insights into the likelihood of synergistic and antagonistic effects. The advantage of increased environmental realism is also of critical importance and will aid in the development of better monitoring programs and regulations. Computer simulations, for instance, based on the chemical mode of action (e.g., Quantitative Structure Activity Relationships (QSARs) see [40]) are an important first step to reduce the time and cost of more detailed assessments while promoting informed decision making.

Joint exposure to multiple stressors has been addressed previously in the ecological literature in the theory of multiple predators (e.g., [11, 41–43]). The multiple predator approach is particularly fitting, as responses to predators are highly variable (e.g., [44]) as are responses to insecticides (e.g., above studies). The theory of multiple predators shows that predator-predator interactions can cause conflicting risk to prey and lays out a framework for assessing the emergent properties of multiple predators on simple food webs. In the ecological framework, each predator is treated as an individual stressor and as such presents an interesting analogy to work with different chemical stressors. The predator framework modified for chemical stressors suggests that there are a series of steps to move forward with cumulative effects risk assessment. These are: (1) to define the criteria for identifying mixtures of likely substances, (2) monitor how common substances interact with each other and environmental compartments, (3) assess what mechanisms may underlie unexpected interactions, and (4) propose how the impacts of multiple stressors on stream communities may be regulated. This approach is far simpler than some of the chemical-based approaches suggested by others while also enabling the inclusion of insights gained using these methods [45, 46]. Finally, a simple, holistic approach that integrates ecological components will likely present a fresh perspective enabling the capture of the complexity of both the mixtures of chemicals under investigation and the interacting assemblages of organisms in real ecosystems.

5. Conclusions

Complex mixtures of five or more pesticides, as well as nutrients and sediments, are pervasive in the aquatic environment. Yet, mortal endpoints of single chemicals on single species laboratory tests are the norm in regulatory frameworks. A more holistic approach is needed. Within the regulatory community, there is a concern that multiple stressor studies are difficult to interpret and as a result, are often ignored. The above synthesis and review of seven mesocosm studies on the combined effects of pesticides, nutrients, and macroinvertebrate community dynamics show that interactions between chemical substances, nutrient enrichment, and trophic status can change how communities respond to stress. This work offers unique insights into the evaluation of multiple stressors as it shows that expected toxic mechanisms can be muted or intensified in response to changing natural and anthropogenic gradients. This finding of diverse responses to stress is consistent with findings from field studies in the literature where some communities tend to be more resilient to stress than others. Understanding multiple stressor effects in an ecological framework (e.g., theory of multiple predators) within a regulatory context may offer a simple and more holistic approach to environmental risk assessment integrating findings from mixture theory and community-level responses to multiple stressors.

Acknowledgements

Dave Hryn and Jon Bailey provided technical expertise in conducting the mesocosm experiments and chemical analyses respectively. Eric Luiker helped with the design and logistics of many of the experiments and Kristie Heard helped with the macroinvertebrate subsampling schema and identifications. This review and synthesis was inspired by feedback from Drs. J.M. Culp, D.J. Baird and Ms. M. MacGregor. Financial support for this research was provided by NSERC (Trusiak PGS-D3 #362641) and a Pest Science Fund grant (Environment and Climate Change Canada).

Conflict of interest

The author declares no conflict of interest.

Author details

Alexa C. Alexander

Address all correspondence to: alexa.alexander@unb.ca

Environment and Climate Change Canada, Department of Biology and Canadian Rivers Institute, University of New Brunswick, Fredericton, New Brunswick, Canada

References

[1] Allan JD. Landscapes and riverscapes: The influence of land use on stream ecosystems. Annual Review Ecology, Evolution and Systematics. 2004;**35**:257-284. DOI: 10.1146/annurev.ecolsys.35.120202.110122

[2] Schulz R. Field studies on exposure, effects, and risk mitigation of aquatic nonpoint-source insecticide pollution: A review. Journal of Environmental Quality. 2004;**33**:419-448. DOI: 10.2134/jeq2004.4190

[3] Gilliom RJ. Pesticides in U.S. streams and groundwater. Environmental Science & Technology. 2007;**41**:3408-3414. DOI: 10.1021/es072351u

[4] Main AR, Headley JV, Peru KM, Michel NL, Cessna AJ, Morrissey CA. Widespread use and frequent detection of neonicotinoid insecticides in wetlands of Canada's prairie pothole region. PLoS One. 2014;**9**:e92821. DOI: 10.1371/journal.pone.0092821

[5] Morrissey CA, Mineau P, Devries JH, Sanchez-Bayo F, Liess M, Cavallaro MC, Liber K. Neonicotinoid contamination of global surface waters and associated risk to aquatic invertebrates: A review. Environment International. 2015;**74**:291-303. DOI: 10.1016/j.envint.2014.10.024

[6] Klarich KL, Pflug NC, DeWald EM, Hladik ML, Kolpin DW, Cwiertny DM, LeFevre GH. Occurrence of neonicotinoid insecticides in finished drinking water and fate during drinking water treatment. Environmental Science & Technology Letters. 2017;**4**:168-173. DOI: 10.1021/acs.estlett.7b00081

[7] Tomizawa M, Casida JE. Selective toxicity of neonicotinoids attributable to specificity of insect and mammalian nicotinic receptors. Annual Review of Entomology. 2003;**48**: 339-364. DOI: 10.1146/annurev.ento.48.091801.112731

[8] Henry M, Béguin M, Requier F, Rollin O, Odoux J-F, Aupinel P, Aptel J, Tchamitchian S, Decourtye A. A common pesticide decreases foraging success and survival in honey bees. Science. 2012;**336**:348-350. DOI: 10.1126/science.1215039

[9] Alexander AC, Culp JM, Liber K, Cessna AJ. Effects of insecticide exposure on feeding inhibition in mayflies and oligochaetes. Environmental Toxicology and Chemistry. 2007;**26**:1726-1732. DOI: 10.1897/07-015R.1

[10] Alexander AC, Heard KS, Culp JM. Emergent body size of mayfly survivors. Freshwater Biology. 2008;**53**:171-180. DOI: 10.1111/j.1365-2427.2007.01880.x

[11] Werner EE, Peacor SD. A review of trait-mediated indirect interactions in ecological communities. Ecology. 2003;**84**:1083-1100. DOI: 10.1890/0012-9658(2003)084[1083,ARO TII]2.0.CO;2

[12] Kramer PRG, Jonkers DA, van Liere L. Interactions of Nutrients and Toxicants in the Food Chain of Aquatic Ecosystems. RIVM Report No. 703715001. Bilthoven, The Netherlands: National Institute of Public Health and the Environment; 1997. 113 p

[13] De Zwart D, Posthuma L. Complex mixture toxicity for single and multiple species: Proposed methodologies. Environmental Toxicology & Chemistry. 2005;**24**:2665-2676. DOI: 10.1897/04-639R.1

[14] Schwarzenbach RP, Escher BI, Fenner K, Hofstetter TB, Johnson CA, von Gunten U, Wehrli B. The challenge of micropollutants in aquatic systems. Science. 2006;**313**:1072-1077. DOI: 10.1126/science.1127291

[15]   Goussen B, Price OR, Rendal C, Ashauer R. Integrated presentation of ecological risk from multiple stressors. Scientific Reports. 2016;**6**:36004. DOI: 10.1038/srep36004

[16]   Kienzler A, Bopp SK, van der Linden S, Berggren E, Worth A. Regulatory assessment of chemical mixtures: Requirements, current approaches and future perspectives. Regulatory Toxicology & Pharmacology. 2016;**80**:321-334. DOI: 10.1016/j.yrtph.2016.05.020

[17]   Hynes HBN. The Ecology of Running Waters. Toronto: University of Toronto Press; 1970. 555 p. DOI: 10.4319/lo.1971.16.3.0593

[18]   Carter JL, Resh VH, Hannaford MJ. Macroinvertebrates as biotic indicators of environmental quality. In: Hauer FR, Lamberti GA, editors. Methods in Stream Ecology, Volume 2: Ecosystem Function. Burlington: Academic Press; 2017. pp. 293-318. DOI: 10.1016/B978-0-12-813047-6.00016-4

[19]   Merritt RW, Cummins KW. An Introduction to the Aquatic Insects of North America. 3rd ed. Kendall Hunt: Dubuque; 1996. 862 p. DOI: 10.2307/1467288

[20]   Edmunds GF, Jensen SL, Berner L. The Mayflies of North and Central America. Don Mills: Burns & MacEachern Ltd.; 1976. 330 p

[21]   Alexander AC, Luis AT, Culp JM, Baird DJ, Cessna AJ. Can nutrients mask community responses to insecticide mixtures? Ecotoxicology. 2013;**22**:1085-1100. DOI: 10.1007/s10646-013-1096-3

[22]   Allan JD. Feeding habits and prey consumption of three setipalpian stoneflies (Plecoptera) in a mountain stream. Ecology. 1982;**63**:26-34. DOI: 10.2307/1937027

[23]   Westfall MJ, Tennessen KJ. Chapter 12, Odonata. In: Merrit RW, Cummins KW, editors. An Introduction to the Aquatic Insects of North America. 3rd ed. Dubuque: Kendall/Hunt Publishing Co.; 1996. pp. 164-211. DOI: 10.2307/1467288

[24]   Cooper SD, Smith DW, Bence JR. Prey selection by freshwater predators with different foraging strategies. Canadian Journal of Fisheries and Aquatic Sciences. 1985;**42**:1720-1732. DOI: 10.1139/f85-216

[25]   Peckarsky BL. Aquatic insect predator-prey relations. Bioscience. 1982;**32**:261-266. DOI: 10.2307/1308532

[26]   Baekken T, Aanes KJ. Pesticides in Norwegian agriculture. Their effects on benthic fauna in lotic environments. Preliminary results. In: Internationale Vereinigung für Theoretische und Angewandte Limnologie: Verhandlungen. 1991;**24**:2277-2281. DOI: 10.1080/03680770.1989.11899943

[27]   van Wijngaarden R, Leeuwangh P, Lucassen WGH, Romijn K, Ronday R, van der Velde R, Willigenburg W. Acute toxicity of chlorpyrifos to fish, a newt, and aquatic invertebrates. Bulletin of Environmental Contamination and Toxicology. 1993;**51**:716-723. DOI: 10.1007/BF00201650

[28] Dunn A. A Relative Risk Ranking of Pesticides Used in Prince Edward Island. Surveillance Report EPS-5-AR-04-03. Environment Canada Environmental Protection Branch: Dartmouth; 2004. 41 p

[29] Health Canada. Proposed Re-evaluation Decision PRVD2011-14 Chlorothalonil. 1 November 2011 [Internet]. 2011. Available from: http://publications.gc.ca/collections/collection_2012/sc-hc/H113-27-2011-14-eng.pdf [Accessed: 2017-12-13]

[30] Pestana JLT, Alexander AC, Culp JM, Baird DJ, Soares AMVM. Structural and functional responses of benthic invertebrates to imidacloprid in outdoor stream mesocosms. Environmental Pollution. 2009;**157**:2328-2334. DOI: 10.1016/j.envpol.2009.03.027

[31] Alexander AC, Culp JM. Predicting the effects of insecticide mixtures on non-target aquatic communities. In: Trdan S, editor. Insecticides—Development of Safer and more Effective Technologies. InTech Online; 2013. pp. 83-101. DOI: 10.5772/3356

[32] Alexander AC, Culp JM, Baird DJ, Cessna AJ. Nutrient-contaminant interactions decouple density-density responses in aquatic and emergent insects. Freshwater Biology. 2016;**61**:2090-2101. DOI: 10.1111/fwb.12711

[33] Biggs BJF. Eutrophication of streams and rivers: Dissolved nutrient-chlorophyll relationships for benthic algae. Journal of the North American Benthological Society. 2000;**19**:17-31. DOI: 10.2307/1468279

[34] Dodds WK, Jones JR, Welch EB. Suggested classification of stream trophic state: Distributions of temperate stream types by chlorophyll, total nitrogen, and phosphorus. Water Research. 1998;**32**:1455-1462. DOI: 10.1016/S0043-1354(97)00370-9

[35] Winer BJ, Brown DR, Michels KM. Statistical Principles in Experimental Design. 3rd ed. New York: McGraw Hill; 1991. 928 p

[36] Clarke KR, Warwick RM. Change in Marine Communities: An Approach to Statistical Analysis and Interpretation. 2nd ed. Plymouth Marine Laboratory: Plymouth; 2001. 175 p

[37] Underwood AJ. Experiments in Ecology: Their Logical Design and Interpretation Using Analysis of Variance. 6th ed. New York: Cambridge University Press; 2002. 504 p. DOI: 10.1017/CBO9780511806407

[38] Strachan SA, Reynoldson TB. Performance of the standard CABIN method: Comparison of BEAST models and error rates to detect simulated degradation from multiple data sets. Freshwater Science. 2014;**33**:1225-1237. DOI: 10.1086/678948

[39] Shipley B. Cause and Correlation in Biology: A user's Guide to Path Analysis, Structural Equations and Causal Inference. Cambridge: Cambridge University Press; 2000. 332 p. DOI: 10.1017/CBO9780511605949

[40] Escher BI, Hermens JLM. Modes of action in ecotoxicology: Their role in body burdens, species sensitivity, QSARs, and mixture effects. Environmental Science & Technology. 2002;**36**:4201-4217. DOI: 10.1021/es015848h

[41] Sih A, Englund G, Wooster D. Emergent impacts of multiple predators on prey. Trends in Ecology & Evolution. 1998;**13**:350-355. DOI: 10.1016/S0169-5347(98)01437-2

[42] Rohr JR, Kerby JL, Sih A. Community ecology as a framework for predicting contaminant effects. Trends in Ecology & Evolution. 2006;**21**:606-613. DOI: 10.1016/j.tree.2006.07.002

[43] Northfield TD, Barton BT, Schmitz OJ. A spatial theory for emergent multiple predator-prey interactions in food webs. Ecology & Evolution. 2017;**7**:6935-6948. DOI: 10.1002/ece3.3250

[44] Wooster D. Predator impacts on stream benthic prey. Oecologia. 1994;**99**:7-15. DOI: 10.1007/BF00317077

[45] Jonker MJ, Svendsen C, Bedaux JJM, Bongers M, Kammenga JE. Significance testing of synergistic/antagonistic, dose level-dependent, or dose ratio-dependent effects in mixture dose-response analysis. Environmental Toxicology & Chemistry. 2005;**24**:2701-2713. DOI: 10.1897/04-431R.1

[46] Altenburger R, Backhaus T, Boedeker W, Faust M, Scholze M, Grimme LH. Predictability of the toxicity of multiple chemical mixtures to *Vibrio fischeri*: Mixtures composed of similarly acting chemicals. Environmental Toxicology & Chemistry. 2000;**19**:2341-2347. DOI: 10.1002/etc.5620190926

Development of Beekeeping: An Analysis Using the Technique of Principal Components

Emerson Dechechi Chambó,
Regina Conceição Garcia, Fernando Cunha,
Carlos Alfredo Lopes de Carvalho,
Daiane de Jesus Oliveira,
Maiara Janine Machado Caldas,
Nardel Luiz Soares da Silva, Ludimilla Ronqui,
Claudio da Silva Júnior, Pedro da Rosa Santos and
Vagner de Alencar Arnaut de Toledo

Additional information is available at the end of the chapter

http://dx.doi.org/10.5772/intechopen.74223

Abstract

Beekeeping is an economic activity of the Brazilian agricultural sector and a powerful tool to achieve sustainable development. However, beekeeping still remains a modest activity compared to other areas, with a lack of technical knowledge and beekeeping practices that need to be standardized. This study represents a proposal for the diagnosis of beekeeping, to facilitate decision-making and to provide a faster development of the beekeeping activity. We investigated the process of adoption of beekeeping practices of 28 beekeepers and the quality of the honey produced by them in the Western region of Paraná, using the technique of Principal Components Analysis after the construction of apicultural indexes. Specifically, the honey produced in the Western region of Paraná be included in the requirements of national and international legislation, but the beekeeping practices adopted still require standardization so that the beekeepers have higher honey production. Also, the transformation of variables into apicultural indexes for later use in the analysis of principal components proved to be efficient to draw a beekeeping profile. Our research proves to be efficient in accurately diagnosing beekeeping bottlenecks, which may enable better decision-making and thus attract new entrepreneurs and increase their relevance to achieve sustainable rural development.

Keywords: multivariate analysis, *Apis mellifera*, physicochemical analyses, apicultural indexes, sustainable rural development

1. Introduction

Beekeeping is an economic activity of the agricultural sector and an environmentally sustainable production model, crucial for biodiversity and agriculture [1]. Beekeeping provides additional income for many poor communities, creating new opportunities in rural areas, and improves the living conditions of many families [2, 3]. In addition, the pollination service provided by honeybees generates increases in crop yields [4] and contributes to the balance of the ecosystem and biodiversity [5].

In the economic sphere, Brazil stands out among the honey-producing countries worldwide, occupying the 8th position. In 2014, Brazil exported 25.317 tons of honey, generating exchange revenue of US $ 98.58 million [6]. In the Brazilian scenario, Rio Grande do Sul, with a honey production of 7.286 tons, is the first State in the national ranking, followed by the States of Paraná with 5.565 tons and Santa Catarina with 4.887 tons of honey produced [7].

Although economic indicators of beekeeping show progress in the activity, its development is far below that expected for a country that has one of the largest biodiversity on the planet and conditions conducive to the beekeeping. Thus, making beekeeping more profitable, such as adopting appropriate beekeeping practices, with a consequent higher honey quality, can be an alternative to attract new entrepreneurs and increase their relevance as a means to achieve sustainable rural development. The physicochemical composition of honey is complex and can be used to verify the quality of the honey produced, as well as to improve honey extraction practices [8, 9], conservation, and storage, avoiding contamination, honey differentiation of other products, and determination of botanical origin [10].

A reliable statistical analysis, correlating information on beekeeping practices and quantity and quality of honey produced in the Western region is very important for the growth of beekeeping since in addition to the aspect of legislation these analyses can support assessments of the beekeeping profile, against many changes in the region in the last decade, thanks to the partnerships between the university and other institutions with beekeepers.

Principal Component Analysis (PCA) is a very powerful technique, but it has its limitations. It is generally used to study variance and covariance through linear combinations of ϱ variables involved [11] and also to order data based on quantitative variables, with or without transformation [12]. PCA was initially defined for data with multinomial distributions [12], although it can be applied to binary data [13]. The objective of this study is to investigate the process of adoption of beekeeping practices and the quality of honey produced in the Western region of Paraná, proposing an alternative analysis of the data through the technique of Principal Component Analysis, using apicultural indexes constructed from binary variables and associated with quantitative variables.

2. Material and methods

2.1. Evaluation questionnaire

The study was carried out from January 2015 to June 2015 with 28 cooperative beekeepers from the Cooperativa Agrofamiliar Solidária dos Apicultores of the West Coast of Paraná, located in the county of Santa Helena, State of Paraná, Brazil (latitude 24°51′37″S and longitude 54°19′58″W).

For data collection, a questionnaire adapted from [14] was applied. The questionnaire consisted of 46 closed-ended questions, in which the response by beekeepers allowed the construction of a binary data matrix (1 or 0), and 4 open-ended questions (quantitative): honey productivity in kg per colony, number of colonies of *Apis mellifera*, age and experience in beekeeping activity in years. Each set of strictly related binary variables (closed-ended questions) that resulted in the construction of the indexes associated with the four beekeeping practices: equipment use, beekeeping management, harvesting and post-harvesting, and management and marketing can be seen in **Table 1**. As the study did not disclose any personal information about its participants, no approval was sought by an ethics committee.

2.2. Construction of apicultural indexes

The construction of the indexes occurred to facilitate the understanding of the set of attributes that determine each important aspect of the apicultural chain, as well as to be analyzed in association with the quantitative variables (open-ended questions) through the Principal Component Analysis.

The construction of the indexes was based on Miranda's method [15] for the determination of technological indexes. The index for each beekeeper j in apicultural practice g ($I_{g(j)}$) is given by:

$$I_{g(j)} = \sum_{i=1}^{n} \frac{x_i}{n} \tag{1}$$

where x_i represents the value [0 or 1] assumed by the i-th variable (with i varying from 1 to N) of the g-th beekeeping practice (with g varying from 1 to N beekeeping practices) in the j-th beekeeper (with j varying from 1 to N beekeepers). The n is the number of variables measured within the specified beekeeping practice.

Thus, for apicultural practice regarding the use of equipment, $g = 1$, $n = 12$, and $i = 1, ..., 12$; for management, $g = 2$, $n = 19$, and $i = 1, ..., 19$; for harvesting and post harvesting, $g = 3$, $n = 8$ and $i = 1, ..., 8$ and for management and marketing, $g = 4$, $n = 7$ and $i = 1, ..., 7$. Therefore, $n = $ Max $\sum_{i=1}^{n} xi$, therefore, $0 \leq I_{jg} \leq 1$.

The mean general index of beekeepers for beekeeping management practices (IMg) is given by the sum of a beekeeping practice divided by the number of beekeepers and is calculated as:

$$IM_g = \frac{1}{N} \sum_{j=1}^{N} \sum_{i=1}^{n} \frac{x_i}{n} = \frac{1}{N} \sum_{j=1}^{N} I_{g(j)} \tag{2}$$

where N is the number of beekeepers measured (N = 28).

The general index for each beekeeper (IPj), including all beekeeping management practices is given by:

$$`IP_j = \frac{1}{g} \sum_{g=1}^{p} I_{g(j)} \tag{3}$$

where g is the number of beekeeping practices.

Beekeeping practices regarding the use of equipment ($n_1 = 12$)

1. Clothing (mask, hat, beekeeping suit, gloves and boots)	(1) use all	(0) some
2. Smoker (fuel)	(1) vegetable origin	(0) animal origin
3. Hive tool	(1) use	(0) not use
4. Bee brush	(1) use	(0) no use
5. Langstroth Hive	(1) standard	(0) not standard
6. Stainless steel equipment	(1) all	(0) some
7. Centrifuge	(1) electric	(0) manual
8. Decanter	(1) use	(0) not use
9. Uncapping table	(1) use	(0) not use
10. Strainer	(1) use	(0) not use
11. Decrystalizer	(1) use	(0) not use
12. Queen excluder	(1) use	(0) not use

Beekeeping practices regarding to management ($n_2 = 19$)

1. Food supply	(1) provide	(0) not provide
2. Queen replacement	(1) yes	(0) no
3. Honeycomb wax replacement	(1) annually	(0) no
4. Supersedure control	(1) use	(0) no
5. Colony division	(1) do	(0) do not
6. Comb management	(1) do	(0) do not
7. Opening of storage space	(1) yes	(0) no
8. Food storage	(1) deep super and honey super	(0) deep super only
9. Fight moths and ants	(1) yes	(0) no
10. Supplemental feeding	(1) provide	(0) not provide
11. Ventilation	(1) use	(0) not use
12. Shading	(1) natural	(0) artificial
13. Distance from water source	(1) < 500 m	(0) > 500 m
14. Uses more than one honey super per colony	(1) yes	(0) no
15. There is honeybee pasture	(1) yes	(0) no
16. Farthest honeybee pasture	(1) < 10 km	(0) > 10 km
17. Minimum proximity of other apiaries	(1) > 3 km	(0) < 3 km
18. Weekly frequency of visits to the apiary	(1) > 1 visit	(0) < 1 visit
19. Rent honeybee pasture	(1) yes	(0) no

Beekeeping practices regarding to honey harvest and post-harvest ($n_3 = 8$)

1. Smoke	(1) use	(0) not use
2. Comb cleaning	(1) yes	(0) no
3. Uncapping fork	(1) use	(0) not use

4. Honey processing	(1) honey house	(0) other
5. Honey super transport	(1) suitable	(0) no
6. Equipment hygiene	(1) yes	(0) no
7. Honey storage	(1) protected	(0) not protected
8. Containers for honey storage	(1) standard	(0) not standard
Beekeeping practices regarding to management and honey marketing ($n_4 = 7$)		
1. Contract for services	(1) yes	(0) no
2. Information on market trends	(1) yes	(0) no
3. Employees training	(1) yes	(0) no
4. Quality control	(1) yes	(0) no
5. Partnership: research	(1) yes	(0) no
6. Marketing	(1) use	(0) not use
7. Computing	(1) use	(0) not use

Table 1. Binary variables related to beekeeping practices.

The general index of the beekeeping system (IA), considering all beekeepers and all beekeeping practices, can be expressed as follows:

$$IA = \frac{1}{N} \sum_{j=1}^{N} IP_j = \frac{1}{8} \sum_{g=1}^{v} I M_g \qquad (4)$$

2.3. Physicochemical analyses of honey

For the physicochemical analysis of honey, 28 honey samples from *Apis mellifera* were collected directly from beekeepers in the Western region of Paraná. Analyses of the physicochemical parameters of the honey samples were performed according to the Official Methods of Analysis, reported in detail in [16]. The evaluated parameters were: moisture (%), ash content (%), electrical conductivity (μS.cm^{-1}), hydroxymethylfurfural content (HMF) (mg.kg^{-1}), acidity (meq.kg^{-1}), diastase activity (Gocthe degrees), reducing sugars (%), apparent sucrose (%) and pH. The analyses were performed in triplicates from each parameter of the sample to obtain the data reported.

2.4. Statistical analyses

Minimum and maximum values, median, 5 and 95% percentiles, mean and standard mean error (SME), and Shapiro-Wilk normality test were calculated for the variables analyzed. Data from the questionnaire were analyzed using the technique of Principal Component Analysis (PCA) after normalization of the data, using Pearson X'X correlation matrix. The multicollinearity diagnosis was performed for the correlation matrix [17]. Subsequently, another PCA was performed without transformation for the sample data related to the physicochemical honey analyses. The Kaiser-Guttman criterion was used to select the number of interpretable axes in the PCA [12]. All statistical analyses were performed using the "R" software version 3.0.2 [18].

3. Results and discussion

Honey production of the beekeepers varied from 15.00 to 40.00 kg.colony^{-1} (mean ± SME of 23.86 ± 1.16 kg.colony^{-1}) and 90% of beekeepers had honey production from 16.80 to 35.00 kg. colony^{-1}. The number of colonies kept by beekeepers varied from 12 to 430 (mean ± SME of 110 ± 19), and only one beekeeper had beekeeping as the main source of income. The age of beekeepers ranged from 26 to 77 years, while experience in the activity ranged from 6 to 50 years, with an approximate average of 22 years of experience in beekeeping (**Table 2**). Beekeeping in the Western region of Paraná is predominantly family friendly.

Camargo et al. [19] developed a Geographic Information System for beekeeping in the Western region of Paraná and found that 46% of the beekeepers had from 5 to 20 colonies, with honey production per colony no larger than 22.12 kg in the larger producer groups (i.e., they had larger areas of forest and smaller areas under agriculture). From our results, in general, it can be seen that the number of colonies kept by beekeepers increased, as well as there was a small increase in honey production per colony compared to 2009.

Possibly, the growth of the legal reserve area on these properties has been one of the factors responsible for such an increase in honey production, since we verified that no beekeepers pay rent to use an apiculture pasture and that good beekeeping practices related to management are not widely adopted (see results below). Brodschneider and Crailsheim [20] reported higher productivity of a colony is linked to the provision of balanced macronutrient nutrition. The increase in the area of forest near apiaries provides an increase in the diversity of

Indexes	Higher	Lower	Median	5%	95%	Mean[1]	SME
Beekeeping equipment	0.92	0.33	0.63	0.33	0.83	0.61	0.03
Beekeeping management practices	0.68	0.21	0.47	0.32	0.66	0.49	0.02
Management and marketing	0.57	0.29	0.29	0.29	0.43	0.33	0.01
Harvesting and post-harvesting techniques	0.90	0.44	0.74	0.64	0.90	0.75	0.02
General index for each beekeeper	0.72	0.34	0.54	0.42	0.68	0.55	0.02
Honey productivity in kg per colony	40.00	15.00	20.50	16.75	35.00	23.86	1.16
Number of colonies	430.00	12.00	75.00	21.75	302.00	110.32	19.39
Experience in beekeeping activity in years	50.00	6.00	19.00	7.35	40.00	21.93	2.22
Age	77.00	26.00	57.00	38.35	65.95	54.68	1.98

[1]Mean obtained from 28 observations.

Table 2. Numerical summary of the survey on the adoption of beekeeping practices.

plants, which positively affects the nutrition of the honeybees and, consequently, increases the colony productivity [21].

The general index of beekeeping (IA) in the Western region of Paraná was 0.55. The general index for each beekeeper, which includes all beekeeping practices ranged from 0.34 to 0.72430 (mean ± SME of 0.55 ± 0.02). Average rates for each beekeeping practice were: use of beekeeping equipment 0.61, management 0.49, harvesting and post-harvesting techniques 0.75, and management and marketing 0.33. There was a great variation of values for each index (see **Table 2**), which indicates the different adoption of practices in beekeeping.

The higher honey production is linked to the adoption of recommended beekeeping practices, especially those related to management (see **Figure 1**). However, only 50% of the beekeepers use more than half of the beekeeping practices recommended for management (median 0.48) (**Table 2**), with 50% of the beekeepers not adopting 50% of management recommendations regarding good beekeeping practices. Obviously, this can be an obstacle to production, because in addition to the honeybee flora, queen and old combs replacement, fighting diseases, and food supply are prime factors for the increase of honey production.

More than 50% of beekeepers adopt more than 74% of the recommendations of beekeeping practices regarding the harvesting and post-harvesting of honey (median = 0.74, **Table 2**). However, it represents the quality of the honey samples verified through the parameters (see **Table 3**), in which only some samples did not include the standards established by national and international legislation [22, 23].

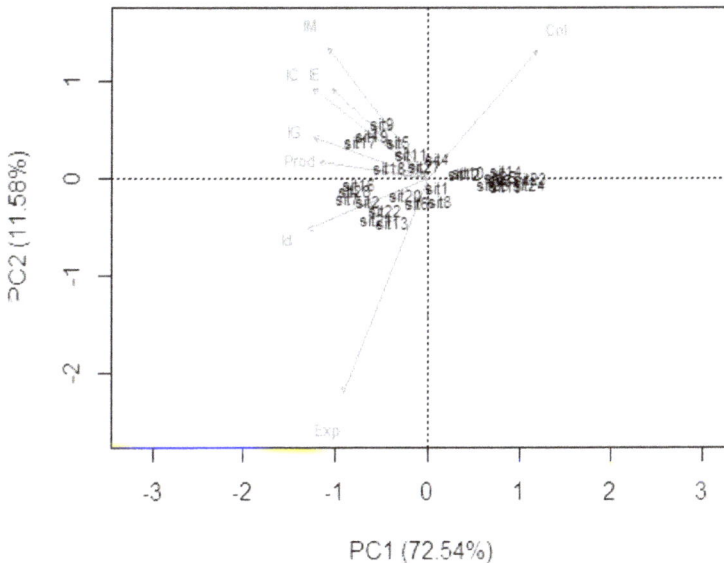

Figure 1. Ordering of the questionnaire data (normalized) in the first two main axes. Beekeeping equipment use (IE), beekeeping management practices (IM), harvesting and post-harvesting (IC), management and marketing (IG), honey productivity in kg per colony (Prod), number of colonies of *Apis mellifera* (Col), age (Id) and experience in beekeeping activity in years (Exp).

Parameters	Higher	Lower	Median	5%	95%	Mean[1]	SME
Acidity (mq.kg^{-1})	49.50	12.75	23.25	13.23	48.16	27.58	2.28
Reducing sugar (%)	90.40	64.42	74.08	66.01	84.92	74.47	1.03
Ash (%)	0.29	0.07	0.12	0.07	0.25	0.13	0.01
Electrical conductivity (μS.cm^{-1})	534.4	120.4	210.4	125.57	388.18	221.67	16.88
Diastase (° Goethe)	21.99	6.02	11.87	6.89	17.65	11.62	0.64
HMF* (mg.kg^{-1})	37.43	0.38	10.26	1.50	21.42	10.04	1.48
pH	4.48	3.44	3.94	3.62	4.30	3.94	0.05
Sucrose (%)	7.37	6.02	3.09	0.98	6.19	3.21	0.33
Moisture (%)	30.05	13.75	18.15	14.78	25.51	19.64	0.77

*Hydroxymethylfurfural.
[1]Mean obtained from 28 observations.

Table 3. Numerical summary of physicochemical parameters of honey samples from *Apis mellifera* colonies, Paraná, Brazil.

The proportion of variance explained by the first two main components for the questionnaire data was 84.12% with the first axis explaining 72.54% and the second main axis 11.58% of the total variation, which in turn is satisfactory to explain most of the variation in PCA [12], see **Figure 1**.

Figure 1 represents a left-to-right gradient, starting with a similar group of the beekeepers (above the X-axis on the left) with higher values for honey productivity in kg per colony, management and marketing, harvesting and post-harvesting techniques, beekeeping equipment and management, which are also more correlated with each other. A second group of the beekeepers (below the X axis on the left) presented higher age and experience in beekeeping, which were variables with high negative correlation with the number of colonies and less positively correlated with honey productivity in kg per colony, management and marketing, harvesting and post-harvesting techniques, beekeeping equipment and management. A third group of beekeepers (to the right of the ordering on the X axis), very similar to each other, presented intermediate values in almost all measured variables.

Table 3 contains the results of the usual descriptive statistics of the eight physicochemical parameters of honey samples analyzed.

Honey acidity varied from 12.75 to 49.50 meq.kg^{-1} (mean ± SME of 27.58 ± 2.28 meq.kg^{-1}), which is in accordance with the requirements of national and international regulations, which is, in general, no more than 50 meq.kg^{-1} [22, 23]. The variation of acidity between the different samples can be attributed to floral origin, harvesting time of honey [8], or fermentation processes [24]. The free acidity of honey can be explained by the presence of organic acids in equilibrium with their corresponding lactones, or internal esters, and some inorganic ions, such as phosphate [25].

Honey is mainly composed of monosaccharides, such as glucose and fructose and disaccharide sucrose. The percentage of reducing sugars of the analyzed honey samples ranged from 65.03 to 90.40% (mean ± SME of 74.47 ± 1.03) and the average percentage of apparent sucrose was 3.21, with a variation of 0.10 to 7.37% and an SME of 0.33. For the percentage of reducing sugars, the Brazilian standard establishes a minimum of 65% [22] and international regulation [23], in general, a minimum of 60%, and all the different analyzed samples was included in these specifications. However, 10% of samples had a percentage of apparent sucrose greater than that required by the legislation which was a maximum of 5% [22, 23].

Electrical conductivity and ash content are important parameters of honey quality [10]. Analyzed honey samples had ash content ranging from 0.07% to 0.29%, that is, below the maximum value of 0.60% [22, 23]. Electrical conductivity values of analyzed honey samples ranged from 120.4 μS.cm^{-1} to 534.4 μS.cm^{-1}. Although there is no value recommended by Brazilian legislation [22] for electrical conductivity, the values obtained are within the scope of international regulation [23], which is desired to be smaller than 800 μS.cm^{-1}. However, in honey samples analyzed from all regions of Brazil, it is very common to obtain values above this index for electrical conductivity.

The ash content is a direct measure of the inorganic residues present in honey sample after the carbonization, while the electrical conductivity measurements express all the organic and inorganic ionizable substances [10]. The electrical conductivity can be considered an important geographical marker for honey samples [26, 27] and ash content may be important in the evaluation of possible mineral contamination [10].

Hydroxymethylfurfural (HMF) and diastase activity are also indicators of honey quality. No sample exceeded the limits set for the HMF parameter, and 13% did not comply with national and international regulations for diastase activity: maximum HMF content of 40 mg.kg^{-1} and a minimum of 8 on the Goethe scale for diastase activity [22, 23]. HMF content of the analyzed honey samples ranged from 0.38 to 37.43 mg.kg^{-1}, and diastase activity ranged from 6.02 to 21.99 on the Goethe scale, indicating that the honey sampled was high quality [10]. It was suggested that for a honey to be considered of high quality it is expected that it has high activity diastase and low content of HMF.

The knowledge of moisture in honey is useful to improve the conservation and storage practices of honey, since it prevents the growth of microorganisms [10]. For the analyzed honey samples, percentage of moisture varied from 13.75 to 30.05, with an average of 19.64% and SME of 0.77. Among the analyzed samples, 33% was not included in the requirements of national and international legislation, which establishes a maximum of 20% of moisture [22, 23]. In the 2008/2009 harvesting, [24] verified that 37.5% of honey samples from the Western region of Paraná presented values higher than 20% of moisture and considered that the responsible factors could be the premature harvesting of honey or the absorption of water from the environment during storage, because it is highly hygroscopic, or because of the amount of rainfall at the time when it was produced.

As honey moisture may be indicative of the mismanagement problem in the region, similar to what occurs in other regions of the country, [24] passed on this information to the regional cooperative beekeepers that provide them technical assistance, so that they are alerted and aware of the correct management. However, it is still apparent from our results that the measurements were not sufficient for the samples to reach the moisture requirements recommended by national and international legislation.

The PCA for the physicochemical parameters (normalized data) indicates that 71.92% of the variation in the data can be explained by the overall effect of the first two main axes. The first two axes have values corresponding to 53.14 and 18.78% of the total variance (**Figure 2**).

In **Figure 2**, PCA data for physicochemical parameters suggest similarities between honey samples, with the formation of only two groups: the first group of samples, to the right of the Y axis (PCA 1), was more similar for acidity, HMF, moisture, diastase, reducing sugars, apparent sucrose, and pH, and a second group, to the left of the Y axis (PCA 2), was more similar for electrical conductivity and ash. **Figure 2** presents a positive correlation between electrical conductivity and ash, acidity, and HMF, between diastase activity and moisture, as well as between reducing sugars, apparent sucrose, and pH.

Therefore, experience in beekeeping, beekeeper age, and a number of colonies kept on properties are not variables that are strongly associated positively with increased honey production. Another issue is that the analyzed honey samples were very similar for the physicochemical parameters, in addition to the ones recommended by national and international legislation, which is an indication of the honey quality of the western region of the State of Paraná, Brazil.

However, we are aware that our results have limitations, such as sample size. Even so, the analyses were efficient and can be a valid alternative for application in future studies. Therefore, new studies must be performed with the same technique of data analysis proposed here, however, with a larger sample size. This could reflect a more representative image of the reality of beekeeping, with the consequence of reducing the main bottlenecks in the honey production chain, aiming at maintaining quality and increasing honey production.

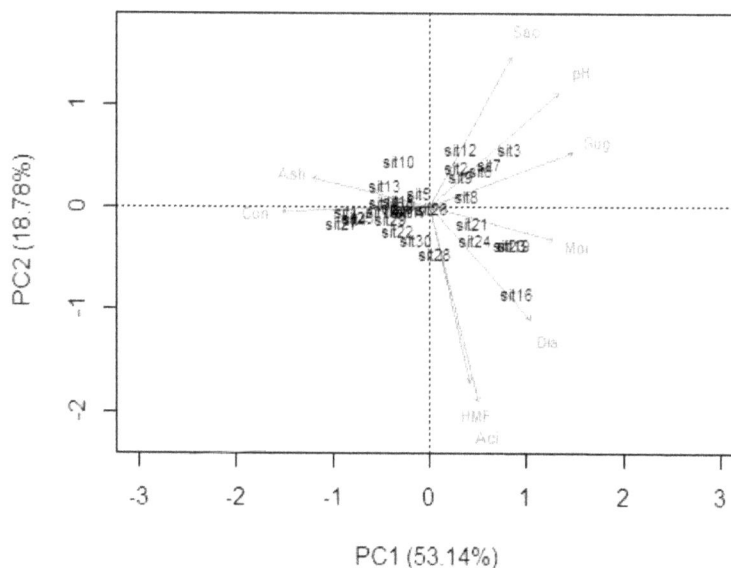

Figure 2. Ordering of the data referring to the physicochemical parameters (normalized) in the plane of the first two main axes. Moisture (Moi), ash content (Ash), electrical conductivity (Con), hydroxymethylfurfural content (HMF), acidity (Aci), diastase activity (Dia), reducing sugars (Sug), and apparent sucrose (Sac).

4. Conclusions

The construction of apicultural indexes and their associations with quantitative variables, using the multivariate technique of Principal Component Analysis was able to explain 84.12% of the total variation of the data in only two main axes and, therefore, proved to be efficient to draw a beekeeping profile, as well as for possible decision-making, with consequences for the future development of the activity. Statistical analysis indicated that the adoption of adequate beekeeping practices in the region, especially those related to the beekeepers' management and marketing of beekeeping products can provide a higher honey production in the region, especially due to its strong positive associations.

Acknowledgements

To CNPq (National Council for Scientific and Technological Development), process number 311663/2014-1 for the financial support, and CAPES (Coordination of Improvement of Higher Level Personnel) for the scholarship for the first author.

Author details

Emerson Dechechi Chambó[1], Regina Conceição Garcia[2], Fernando Cunha[2],
Carlos Alfredo Lopes de Carvalho[3], Daiane de Jesus Oliveira[3],
Maiara Janine Machado Caldas[3], Nardel Luiz Soares da Silva[1], Ludimilla Ronqui[4],
Claudio da Silva Júnior[5], Pedro da Rosa Santos[5] and Vagner de Alencar Arnaut de Toledo[5]*

*Address all correspondence to: vagner_abelha@yahoo.co.uk

1 Nature and Culture Institute, Federal University of Amazonas, Benjamin Constant, Brazil

2 Agrarian Sciences Center, Western of Paraná State University, Marechal Cândido Rondon, Brazil

3 Insecta Research Group, Agrarian Sciences, Environmental, and Biological Center, Federal University of Recôncavo of Bahia, Cruz das Almas, Brazil

4 Federal University of Paraná, Palotina, Brazil

5 Animal Science Department, Maringá State University, Maringá, Brazil

References

[1] Allsopp MH, de Lange WJ, Veldtman R. Valuing insect pollination services with cost of replacement. PLoS One 2008;3:e3128. DOI: 10.1371/journal.pone.0003128

[2] Jaffé R, Pope N, Carvalho AT, Maia UM, Blochtein B, Carvalho CAL, Carvalho-Zilse GA, Freitas BM, Menezes C, Ribeiro MF, Venturieri GC, Imperatriz-Fonseca VL. Bees for development: Brazilian survey reveals how to optimize stingless beekeeping. PLoS One. 2015;**10**:1-21. DOI: 10.1371/journal.pone.0121157

[3] Slaa EJ, Sánchez-Chaves LA, Malagodi-Braga KS, Hofstede FE. Stingless bees in applied pollination: Practice and perspectives. Apidologie. 2006;**37**:293-315. DOI: 10.1051/apido:2006022

[4] Chambó ED, Oliveira NTE, Garcia RC, Duarte-Júnior JB, Ruvolo-Takasusuki MCC, Toledo VAA. Pollination of rapeseed (*Brassica napus*) by Africanized honeybees (Hymenoptera: Apidae) on two sowing dates. Anais da Academia Brasileira de Ciências. 2014;**86**:2087-2100. DOI: 10.1590/0001-3765201420140134

[5] Paxton R. Conserving wild bees. Bee World. 1995;**76**:53-55. DOI: 10.1080/0005772X.1995.11099242

[6] ABEMEL (Associação brasileira dos exportadores de mel). Setor apícola brasileiro em números; 2015. Available in: http://brazilletsbee.com.br. Access September 29, 2015

[7] IBGE (Instituto Brasileiro de Geografia e Estatística). Produção da pecuária municipal; 2013;41. Available in: ftp.ibge.gov.br. Access September 29, 2015

[8] Pérez-Arquillué C, Conchello P, Arinõ A, Juan T, Herrera A. Physicochemical attributes and pollen spectrum of some unifloral Spanish honeys. Food Chemistry. 1995;**54**:167-172. DOI: 10.1016/0308-8146(95)00022-B

[9] Terrab A, Recamales AF, Hernanz D, Heredia FJ. Characterization of Spanish thyme honeys by their physicochemical characteristics and mineral contents. Food Chemistry. 2004;**88**:537-542. DOI: 10.1016/j.foodchem.2004.01.068

[10] Estevinho LM, Féas X, Seijas JÁ, Vázquez-Tato MP. Organic honey from Trás-Os-Montes region (Portugal): Chemical, palynological, microbiological and bioactive compounds characterization. Food and Chemical Toxicology. 2012;**50**:258-264. DOI: 10.1016/j.fct.2011.10.034

[11] Guimarães A. Análise de componentes principais aplicada à estimação de parâmetros no modelo de regressão logística quadrático. Tema. 2013;**14**:57-68. DOI: 10.5540/tema.2013.014.01.0057

[12] Bocard D, Gillet F, Legendre P. Numerical Ecology with R. New York: Springer; 2011. p. 306

[13] Legendre P, Legendre L. Numerical Ecology. 3rd ed. Amsterdam: Elsevier; 2012. p. 1006

[14] Freitas DGF, Khan AS, Silva LMR. Nível tecnológico e rentabilidade de produção de mel de abelha (*Apis mellifera*) no Ceará. Revista de Economia e Sociologia Rural. 2004;**42**:171-188. DOI: 10.1590/S0103-20032004000100009

[15] Miranda EAA. Inovações tecnológicas na viticultura do sub-médio São Francisco (thesis). Recife: Federal University of Pernambuco; 2001

[16] Feás X, Pires J, Iglesias A, Estevinho ML. Characterization of artisanal honey produced on the Northwest of Portugal by melissopalynological and physicochemical data. Food and Chemical Toxicology. 2010;**48**:3462-3470. DOI: 10.1016/j.fct.2010.09.024

[17] Montgomery DC, Peck EA. Introduction to Linear Regression Analysis. New York: John Wiley & Sons; 1981. p. 504

[18] R Core Team. *R:* A language and environment for statistical computing. R Foundation for Statistical Computing, Vienna, Austria; 2017. Available in: http://www.R-project.org/

[19] Camargo SC, Garcia RC, Feiden A, Vasconcelos ES, Pires BG, Hartleben AM, Moraes FJ, Oliveira L, Giasson J, Mittanck ES, Gremaschi JR, Pereira DJ. Implementation of a geographic information system (GIS) for the planning of beekeeping in the west region of Paraná. Anais da Academia Brasileira de Ciências. 2014;**86**:241-258. DOI: 10.1590/0001-3765201420130278

[20] Brodschneider R, Crailsheim K. Nutrition and health in honey bees. Apidologie. 2010;**41**:278-294. DOI: 10.1051/apido/2010012

[21] Brodschneider R, Moosbeckhofer R, Crailsheim K. Surveys as a tool to record winter losses of honey bee colonies – A 2-year case study in Austria and South Tyrol. Journal of Apicultural Research. 2010;**49**:23-30. DOI: 10.3896/IBRA.1.49.1.04

[22] BRASIL (Ministério da Agricultura, Pecuária e Abastecimento). Instrução normativa 11, regulamento técnico de identidade e qualidade do mel; 2000. Available in: http://www. agricultura.gov.br/. Access in: September 29, 2015

[23] Codex Alimentarius. Revised Codex Standard for Honey, Codex STAN 12-1981, Rev. 1 (1987), Rev. 2. 2001

[24] Moraes FJ, Garcia RC, Camargo SC, Pires BG, Hartleben AM, Liesenfeld F, Pereira DJ, Mittanck ES, Giasson J, Gremaschi JR. Caracterização físico-química de amostras de mel de abelhas africanizadas dos municípios de Santa Helena e Terra Roxa (PR). Arquivos Brasilerios de Medicina Veterinária. 2014;**66**:1369-1375. DOI: 10.1590/1678-6865

[25] Finola MS, Lasagno MC, Marioli JM. Microbiological and chemical characterization of honeys from Central Argentina. Food Chemistry. 2007;**100**:1649-1653. DOI: 10.1016/j. foodchem.2015.12.046

[26] Acquarone C, Buera C, Elizalde B. Pattern of pH and electrical conductivity upon honey dilution as a complementary tool for discriminating geographical origin of honeys. Food Chemistry. 2007;**101**:95-703. DOI: 10.1016/j.foodchem.2006.01.058

[27] Sodré GS, Marchini LC, Moreti ACCC, Otsuk IP, Carvalho CAL. Caracterizacão físico-química de amostras de méis de *Apis mellifera* L. (Hymenoptera: Apidae) do Estado do Ceará. Ciencia Rural. 2007;**37**:1139-1144. DOI: 10.1590/S0103-84782007000400036

Chapter 5

Role of Sex Peptide in *Drosophila* Males

Béatrice Denis, Benjamin Morel and
Claude Wicker-Thomas

Additional information is available at the end of the chapter

http://dx.doi.org/ 10.5772/intechopen.74416

Abstract

Drosophila male sex peptide ACP70A is a small peptide mainly produced in the accessory glands. It elicits a high number of post-mating responses in mated females; yet its function in male physiology is not well known. Here, we explore its role in male sex behavior and pheromone biosynthesis, using males either mutant or RNAi knocked-down for *Acp70A*. Courtship was severely affected in both *Acp70A* mutants and *Acp70A* knocked-down males, with only 2% of the males succeeding copulation. Cuticular hydrocarbon amounts were moderately affected with 25% decrease in *sp0* mutant (without *Acp70A* expression) and 10–22% increase in flies overexpressing *Acp70A*. *Acp70A* knock-down either ubiquitously or in the testes surprisingly resulted in an overproduction of hydrocarbons, whose amounts were double of the controls. We tested eight putative "off-target" genes but none of these led to an increase in hydrocarbon amounts. These results show that male courtship behavior is largely dependent on the presence of Acp70A and independent of cuticular hydrocarbons. The presence of potential "off-target" genes explaining the hydrocarbon phenotype is discussed.

Keywords: cuticular hydrocarbons, pheromones, sex peptide, Acp70A, courtship behavior, *Drosophila*

1. Introduction

In most reproducing animals, including *Drosophila*, seminal fluid is transferred along with sperm to females during mating. These seminal fluid components have important effects on female behavior and physiology and have been extensively studied in *Drosophila melanogaster* [1, 2]. Most of these seminal proteins are synthesized in the accessory glands (AGs) and therefore, named ACcessory gland Proteins (ACPs). One well-characterized ACP, ACP70A (also called SP, sex peptide), plays a major role in eliciting postmating response: it modifies the female behavior, resulting in the rejection of courting males [3, 4]. It has a crucial role on female reproduction: it increases oogenesis [5], egg production [6] and egg-laying [3, 4]. It also induces dramatic effects on female nutrition: it increases food uptake [7], modifies food preference by altering nutrient balancing [8] and alters gut water absorption and intestinal transit [9]. The other physiological modifications are the inhibition of sleep [10] and the regulation of sperm release from the storage organs [11]. All these effects are caused by the binding of the C-terminal part of the sex peptide to a neuronal sex peptide receptor (SPR) in the female [12–14]. The central part of sex peptide elicits the expression of immune response genes [15], and the N-terminal part activates the corpora allata (CA), inducing increased synthesis of juvenile hormone (JH) [16], which triggers oogenesis and vitellogenic oocyte progression [5] and also leads to decreased pheromone biosynthesis [17].

Whereas, there are numerous studies on the role of male sex peptide on female physiology, there are no such studies concerning male physiology. As we observed that there was a defect in courtship behavior of sex peptide mutant males, we wanted to elucidate the possible roles of ACP70A in male behavior and physiology. In this study, we report clear defects in male sex behavior and moderate defects in hydrocarbon and pheromone synthesis concerning mutant males. Using sex peptide knocked-down males, we confirmed the control of sex peptide on male sex behavior. Conversely, ubiquitous expression of *Acp70A*-RNAi resulted in a two-fold increase in cuticular hydrocarbon (CHC) amounts. We could exclude the role of eight off-targets in this CHC augmentation and localize this RNAi effect in the accessory glands (responsible for a 35% increase) and in the testes (responsible for the rest of the effect). The presence of sperm in the testes does not affect CHC biosynthesis.

2. Materials and methods

2.1. *Drosophila* strains and rearing

Three strains mutant for sex peptide were used:

- the deficiency *Δ130/TM3* (covering the *Acp70A* gene);

- the point mutant *sp0*, produced by targeted mutagenesis by homologous recombination [4]. *sp0* males were used balanced by *TM3* (*sp0/TM3*: one copy of *Acp70A* is active) or homozygous (*sp0/ sp0*: no production of ACP70A), or crossed by *Δ130/TM3* (*sp0/Δ130*: no production of ACP70A).

- DTA-E [18], which are sperm-less and lack ACPs produced from the main cells (96% of the accessory glands).

The laboratory wild-type Canton-S strain was also used as a control.

The following Gal-lines from the Bloomington *Drosophila* Stock Centre were used: *daughterless (da)-Gal4*, a ubiquitous driver; *elav-Gal4*, a driver expressed in the nervous system [19], *dopa decarboxylase (ddc)-Gal4*, expressed in epidermis and nervous system [20], *1407-Gal4* and

PromE-Gal4, both expressed in pupal and adult oenocytes [21, 22], *c564-Gal4*, expressed in fat body [23], *Acp26A-Gal4*, expressed in accessory glands [3], *svp-Gal80*, which specifically blocks Gal4 activity in the oenocytes [24]. Using a *UAS-GFP* line, we could show that *1407-Gal4* was also expressed in testes and built a line with the following genotype: *1407-Gal4; svp-Gal80* that drives the expression only in the testes. Images were visualized and photographed on a Nikon eclipse E800 microscope with a Cool Snap camera.

A *UAS-Acp70A* line was generated in our laboratory and noted *UAS-Acp70A+* [17]. The following *UAS-RNAi*-lines were obtained from the VDRC Stock Center and directed against: *Acp70A*, SP (109,175 KK); *lamp1*, CG3305, (7309 GD); *dco*, CG4379 (101,524 KK); *rgk1*, CG44011 (108,710 KK); *CG5961* (100,023 KK); *CG15128* (100238KK); *CG9413* (108,867 KK); *CG8315* (105,654 KK); *tinc*, CG31247 (101,175 KK).

Drivers were maintained as heterozygous over a Balancer (Cyo or TM3). In all RNAi knockdown (or overexpression) experiments, balanced gal4-driver females were crossed to UAS males. Balanced progeny was taken as the control of RNAi knocked-down (or overexpression) progeny.

Flies were grown at 25°C with 12/12 light–dark (LD) cycles, on standard cornmeal medium. They were separated by sex at emergence and kept sex-separated in groups of 10 in fresh food vials until testing (4 days after emergence).

2.2. Hydrocarbon analyses

CHCs were removed from single 4-day-old flies by washing them for 5 min in 100 μL heptane containing 500 ng hexacosane (*n*-C26) as an internal standard. The fly was then removed from the vial and 5 μL of each sample was injected into a Perichrom Pr200 gas chromatograph, with hydrogen as the carrier, using a split injector (split ratio 40:1). The oven temperature started at 180°C, ramped at 3°C/min to 300°C, for a total run of 40 min. The data were automatically computed and recorded using Winilab III software (version 04.06, Perichrom) as previously described [17]. As we did not observe significant variation in the CHC profiles, we only represented the total amount of CHCs as means ± SEM (n = 10 for all tests).

2.3. Analysis of *Acp70A* expression

Quantitative PCR was performed as described [25] using RNA TRIzol™ (Invitrogen) to extract RNAs from 10 adults for each sample. cDNAs were synthesized with SuperScript II, and PCR was perfumed with a LightCycler® 480 SYBR Green I Master (Roche Applied Science). Primers for *Acp70A* (5′ ATTCTTGGTTCTCGTTTGCG-3′ and 5′-TAACATCTTCCACCCCAGG-3′) were used. To normalize mRNA amounts, we tested six different genes and used one gene, which was shown to be very stable in all samples: CG7598 (5′-AACGGATGTGGTGTTCGATT-3′ and 5′-TAATGCCATCCTTGGTGTGA-3′). Samples were performed in independent triplicates (each consisting of two technical replicates).

2.4. Mating experiments

A 4-day-old Canton-S female was introduced into the observation chamber, consisting of a watch glass (28-mm diameter and 5-mm internal height) placed on a glass plate and left for

2 min before the introduction of the male. The following parameters were recorded: lengths of courtship, first copulation attempt and copulation latency (time from introduction of the male into the observation chamber to courtship, first copulation attempt or copulation), percentages of courtship, first copulation attempt and copulation (percentages of males performing courtship, copulation attempt or copulation). The effects of genotypes were evaluated by Kruskal-Wallis tests (latencies) and χ^2-tests (percentages of flies). $N \geq 50$ for all tests.

3. Results

3.1. Expression of *Acp70A* in adult males

Acp70A expression was not significantly different in controls (Canton-S and *Acp26A*) and in *sp0* mutants that possess a point mutation in the signal sequence. In contrast, *Acp70A* expression was dramatically inhibited in *Acp26A > Acp70A-RNAi* males (~99%) and higher expression was observed in *Acp26A > Acp70A* males (+63%) (**Figure 1**).

3.2. Effect of *Acp70A* on male sex behavior

Males mutant for *Acp70A* (*sp0/+* and *sp0/sp0*), overexpressing *Acp70A* (*Acp26A > Acp70A*) or RNAi knocked-down (*Acp26A > Acp70A-RNAi* and *da > Acp70A-RNAi*) were tested in face of wild-type females (**Figure 2**).

All the steps of courtship were affected in *sp0* mutants: the number of heterozygous males that attempted or succeeded copulation decreased by 54 and 73%, respectively. The effect of the homozygous mutation was dramatic: *sp0/sp0* males performing courtship (wing vibration) were 5 times fewer than heterozygous or control males, and out of the 50 homozygous

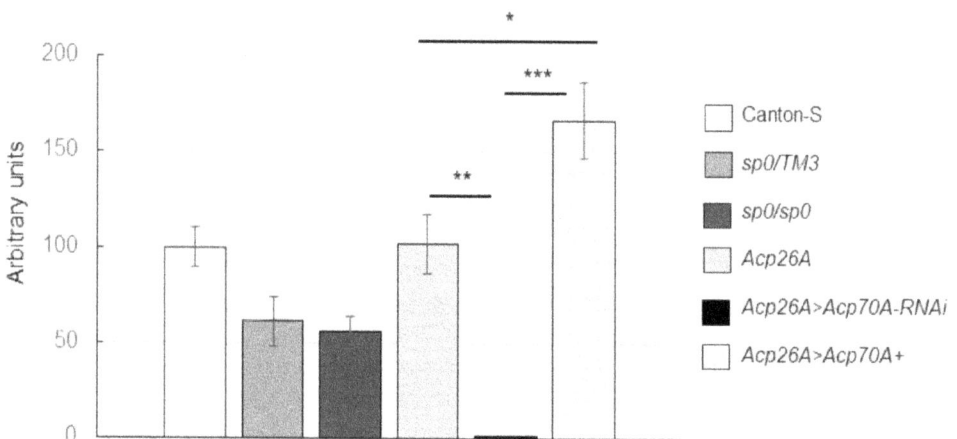

Figure 1. Transcriptional expression of *Acp70A* in control, mutant, knocked-down or overexpressing male flies. Each bar represents mean ± SEM of three independent trials. *, ** and *** indicate significant differences (P = 0.05, 0.01 and 0.001, respectively).

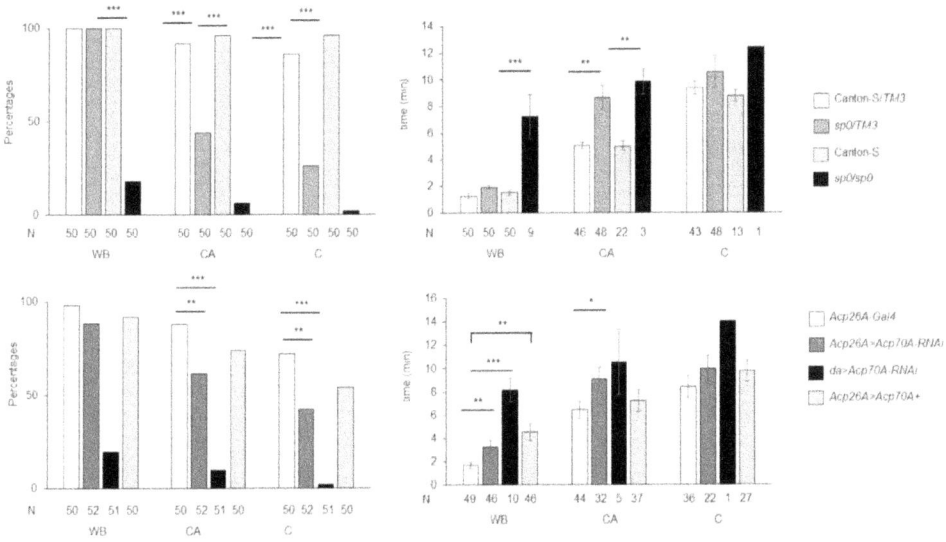

Figure 2. Courtship and mating experiments in fly pairs composed of a wild-type (Canton-S) female and a male of a different genotype: percentages of males performing courtship (WB), copulation attempts (CA) and copulation (C) and time needed to initiate these tasks. Effect of the *sp0* mutation and overexpression or RNAi knock-down of *Acp70A* in males (drivers *Acp26A-Gal4* and *da-Gal4*). Each bar represents mean ± SEM of 50 trials. *, ** and *** indicate significant differences (*P* = 0.05, 0.01 and 0.001, respectively). N is indicated below each bar.

males tested, only 3 attempted to copulate and 1 succeeded copulation. Time needed to perform these tasks was higher as well: homozygous *sp0* males needed 5 times more than *sp0/+* or wild-type males to initiate courtship and the time to attempt copulation was 1.7 and 2 times longer in heterozygous and homozygous mutants, compared to control males.

Overexpression of *Acp70A* in the accessory glands did not modify the proportion of males performing the different steps of courtship behavior. On the other hand, the time necessary to perform wing vibration was double.

Courtship of males knocked-down for *Acp70A* in accessory glands was also affected: the percentage of these males performing copulation attempts and copulation was, respectively, 30 and 41% lower when compared to control males. It took them 2 and 1.5 times longer to perform wing vibration and copulation attempts when compared to control males. When knock-down was induced ubiquitously (*da* > *Acp70A-RNAi*), the inhibition of courtship was more severe and similar to that observed in homozygous *sp0* males.

These results show that there is a significant inhibition of courtship behavior in absence of sex peptide expression.

3.3. Effect of *Acp70A* on CHCs

Heterozygous *sp0* male CHCs were not significantly different from wild-type ones (**Figure 3**). Conversely, homozygous *sp0* males as well as males bearing one *sp0* over a deficiency covering the entire *Acp70A* gene showed a 25% decrease in the total CHC amount. This result

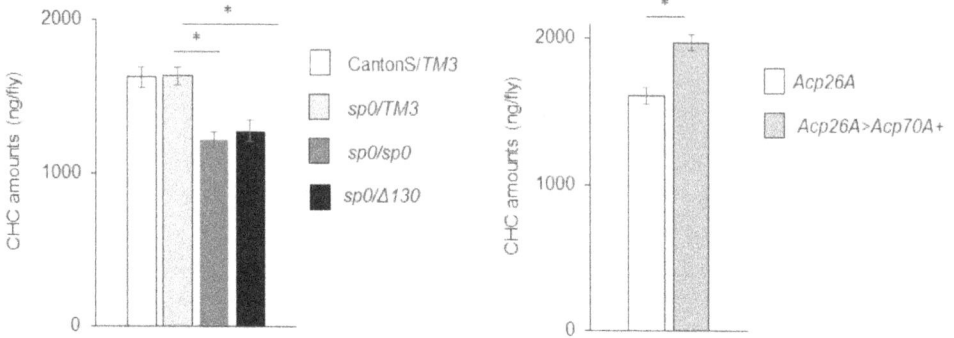

Figure 3. Cuticular hydrocarbon amounts in adult males either mutant for *sp0* (left) or overexpressing *Acp70A* (right) under the *Acp26A-gal4* driver. Each bar represents mean ± SEM (n = 10). * indicates significant differences (*P* = 0.05).

confirms that *sp0* is a null mutant. Inversely, a ubiquitous overexpression of *Acp70A* led to a small but significant increase in CHCs (+22%).

Acp70A ubiquitous knock-down (*da > Acp70A-RNAi*) was followed by a twofold increase in CHC amount (**Figure 4**). We thus wondered whether this increase could be due to off-target effect. The RNAi line was described as having no off-target sequence (no gene covering a 19-mers sequence of the RNAi sequence). We performed a Blast analysis with different 16-mers from the RNAi sequence and obtained eight putative off-target genes containing a stretch of coding sequence identical to at least 15-mers of *Acp70A* sequence (**Figure 5**). The RNAi of these genes was expressed ubiquitously to measure their effect on male CHCs. For five RNAi tested, we obtained no effect on CHCs and for three RNAi (directed against *lamp1*, *rgk1* and *tinc*), there was a lower amount of CHCs (from −14 to 22%, depending on the RNAi) (**Figure 6**). In

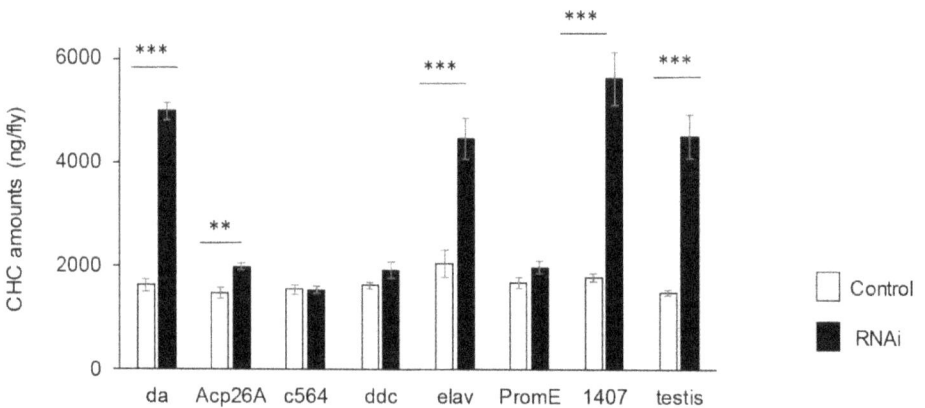

Figure 4. Cuticular hydrocarbon amounts in adult males that were RNAi knocked-down for *Acp70A* in different tissues: ubiquitously (da), in the accessory glands (Acp26A), in fat body (c564), in epidermis (ddc), in nervous system (elav), in oenocytes (Prome), in oenocytes and testes (1407) and in testis (1407; svpgal80). Each bar represents mean ± SEM (n = 10). ** and *** indicate significant differences (*P* = 0.01 and 0.001, respectively).

```
ATGAAAACTCTAGCTCTATTCTTGGTTCTCGTTTGCGTACTCGGCTTGGT    CG5961  CG9413
CCAGGCCTGGGAATGGCCGTGGAATAGGAAGCCTACAAAGTTTCCAATTC    CG8315  CG44011
CAAGCCCCAATCCTCGTGATAAGTGGTGCCGTCTTAATTTGGGGCCCGCC    Lamp1   dco  tinc
TGGGGTGGAAGATGTTAA                                    CG15128
```

Figure 5. Putative off-target genes containing a stretch of coding sequence identical to at least 15-mers of *Acp70A* sequence.

conclusion, the dramatic increase in CHC amount following ubiquitous *Acp70A* knock-down cannot be explained by an off-target effect due to these genes.

3.4. Characterization of the tissue involved in CHC control

In males, *Acp70A* is expressed at a very high level in accessory glands and at a moderate level in testis and carcass (8631, 100 and 95 arbitrary units, respectively; FlyAtlas).

We then wanted to determine the tissue responsible for this effect by targeting *Acp70A-RNAi* to various tissues. We confirmed the locations of expression of the different *Gal4* lines used in this study and showed that *1407-Gal4* was additionally expressed in the testes (**Figure 7**).

No significant effect on CHCs was obtained when *Acp70A RNAi* was expressed in fat body (c564 > *Acp70A RNAi*), in epidermis (*ddc* > *Acp70A RNAi*) and in oenocytes (*PromE* > *Acp70A RNAi*). On the other hand, *Acp70A* knock-down in accessory glands led to a moderate (+35%) increase in CHC amount. CHC amount was multiplied by a factor of 2 *in elav* > *Acp70A RNAi* (nervous system) and a factor of 3 in *1407* > *Acp70A RNAi* (oenocytes + testes) and *1407; svp-Gal80* > *Acp70A -RNAi* (testes). This last result shows an essential role of the testes on CHC production (**Figure 4**).

3.5. CHC profile of the DTAE-line

The DTA-E line is characterized by the absence of accessory glands and some defects in testes, among them, a lack of sperm. DTA-E males were found to produce 1.4-fold more CHCs. We

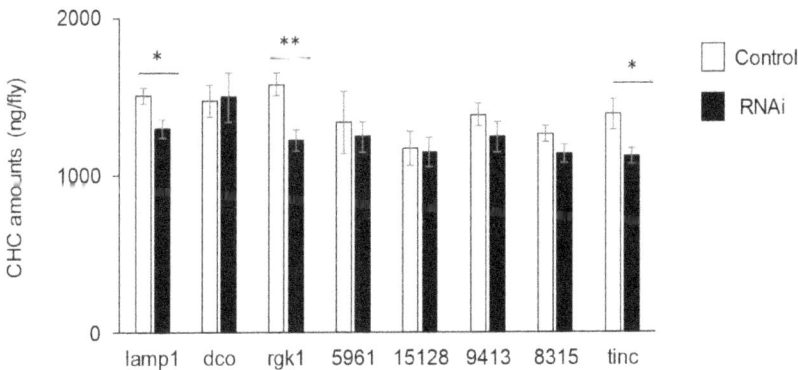

Figure 6. Cuticular hydrocarbon amounts in adult males knocked-down for putative off-target genes. Each bar represents mean ± SEM (n = 10). * and ** indicate significant differences (*P* = 0.05 and 0.01, respectively).

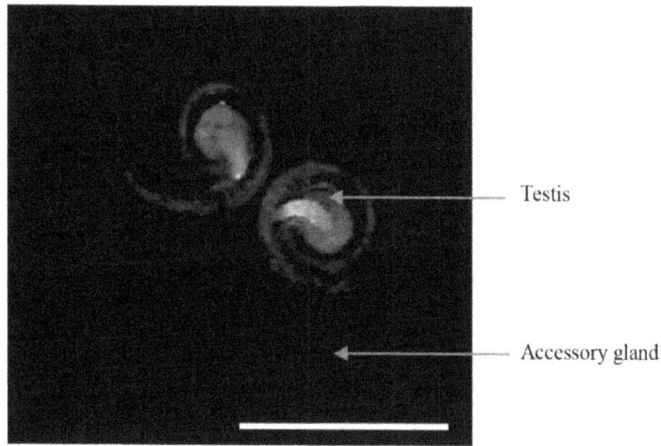

Figure 7. Photomicrographs showing GFP expression in male reproductive apparatus from *1407-Gal4; svp-Gal80*. Fluorescence could be detected only in the testes. Scale bar: 0.5 mm.

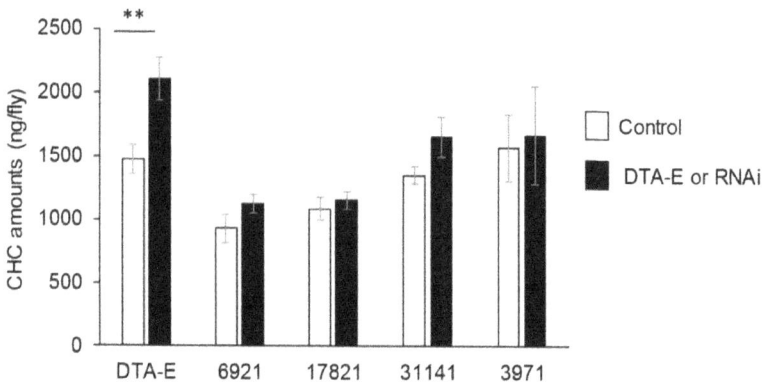

Figure 8. Cuticular hydrocarbon amounts in adult males that do not produce sperm: either DTA-E or knocked-down for CG6821, CG17821, CG31141 and CG3971. Each bar represents mean ± SEM (n = 10). ** indicates significant difference ($P = 0.01$).

wanted to evaluate the effect of the absence of sperm on CHCs. Four elongase genes are essential to spermatozoid development and the lack of expression in testes leads to sterile males without sperm [26–28]. We knocked-down these genes in the testes, using the 1407-Gal4 line. We verified the absence of sperm in the RNAi males. None of these genes had any effect on male CHC production (**Figure 8**).

4. Discussion

4.1. Sex behavior

Ubiquitous overexpression of sex peptide had no significant effect on male sex behavior: the percentage of males performing the different steps of courtship (wing vibration, copulation attempts and copulation) was unchanged and only the time to begin courtship was length-ened. Conversely, *sp0* males showed difficulties to court and the effect was dependent on the dose of the mutant allele: heterozygous *sp0* males courted wild-type females the same way as wild-type males did but only a half of them attempted copulation and one-eighth succeeded to mate. The inhibition was more drastic in homozygous *sp0* males, as less than one-fifth courted the females and only 2% succeeded to mate.

We tested the males that were RNAi knocked-down for sex peptide in the accessory glands. To target the expression in the accessory glands, we used the driver *Acp26A-Gal4*. *Acp26A* gene is almost exclusively expressed in the accessory glands (3589 and 97 units in the accessory glands and the testes, respectively; FlyAtlas). Courtship behavior of *Acp26A > Acp70A RNAi* males was affected, but less than that of *sp0* males: they courted wild-type females the same way as wild-type males did, two-third knocked-down males attempted copulation and less than a half copulated. This result raised the question: does *sp0* affect tissues other than accessory glands? When we ubiquitously expressed sex peptide RNAi, we obtained courtship results similar to those with *sp0*. Taken together, the results suggest a positive control of sex peptide on male courtship behavior. They also pose the problem of the reason of the absence of mat-ing in *sp0* and *da > Acp70A RNAi* males since Q-PCR results clearly show that the expression of *Acp70A RNAi* in accessory glands via *Acp26A-Gal4* reduces *Acp70A* expression to only 1%.

4.2. Cuticular hydrocarbons

In the female, the transfer of ACP70A during mating induces a decrease in cuticular hydrocar-bon amount. This decrease occurs 3 and 4 days after mating and might be due to the overpro-duction of juvenile hormone following mating, caused by the action of *Acp70A* on the corpora allata [17]. We therefore wondered whether *Acp70A* could regulate the production of hydrocar-bons in the male. The *sp0* mutation as well as *Acp70A* ubiquitous overexpression led to mod-erate effects on male CHC production: whereas, wild-type and *sp0* heterozygous males had similar CHC amounts, there was a 25% decrease and a 10–22% increase in homozygous *sp0* that do not produce ACP70A and da > *Acp70A* (overproduction of ACP70A) males, respectively. This result seems to be in favor of a positive regulation of sex peptide on CHC production.

The results concerning the effect of *Acp70A RNAi* on cuticular hydrocarbons were unex-pected: a 35% increase occurred when *Acp70A* expression was inhibited in the accessory glands, using *Acp26A-Gal4*. *Acp26A* gene is mainly, but not exclusively, expressed in the accessory glands (3589 and 97 units in the accessory glands and the testes, respectively; FlyAtlas). *Acp26A* expression in the testes represents 2.7% of the expression in the acces-sory glands, similar to *Acp70A* (1.1%). Moreover, a ubiquitous *Acp70A* knock-down led to a twofold increase in CHC amount; we firstly ascribed this dramatic effect to the presence of possible off-targets of the RNAi.

ACP70A is a small peptide (55 amino acids, including the signal sequence). The nucleic sequence of *Acp70A RNAi* covers almost the totality of the coding sequence, and also includes the small intron. We found eight putative off-target genes, containing a stretch of coding sequence identical to at least 15-mers of *Acp70A RNAi* sequence. However, none of these putative off-target genes could be accountable for the dramatic CHC increase resulting in *Acp70A RNAi* expression.

4.3. Search of the tissue involved in the control on hydrocarbon production

We knocked-down sex peptide in different tissues and could demonstrate that neither the fat body, nor the oenocytes or the epidermis could be responsible for the large rising level of CHCs. On the other hand, sex peptide expression in the testes or in the nervous system led to a CHC increase similar to ubiquitous overexpression.

Sex peptide Acp70A is mainly expressed in the accessory glands, but some expression is also observed in the testes and the carcass (FlyAtlas). Inside the accessory glands, it is exclusively produced by the main cells (96% of the accessory glands) [29]. When we used the DTA-E line in which accessory gland main cell function was genetically disrupted [18], we obtained as well a large-fold increase in CHCs. DTA-E line was obtained after the introduction of diphtheria toxin subunit A (DTA) into the accessory glands via the promoter of *Acp95EF* [18]. ACP95EF is also a sex peptide produced in the accessory glands and transmitted to the female after mating. It has the same place of production as ACP70A; in the accessory glands, it is exclusively produced in the main cells [29]. Within the fly, it is mainly expressed in the accessory glands and marginally in the testes (787 and 62 arbitrary units, respectively; FlyAtlas). DTA-E males lack ACPs produced from the main cells but have normal secondary cells as well as ejaculatory bulb and duct [30]. DTA-E males are sterile and the block of spermatogenesis occurs at the primary spermatocyte stage [18]. The occurrence of a faint expression of this gene in the testes (FlyAtlas) could explain the lack of sperm. However, the lack of sperm is not directly responsible for the large increase in CHC amounts since flies that did not produce sperm after RNAi knock-down for different elongases involved in sperm production did not increase their CHC production.

The question is: why does DTA-E line show a similar male CHC phenotype to *da > Acp70A-RNAi*? In the former line, no off-target can be involved. An explanation could be that a "leakage" of the Acp95EF promoter has resulted in a lack of sperm and probably other defects [18]. In males that have been RNAi knocked-down ubiquitously, one may suppose the effect of unknown "off-target" genes that are essential to testis function. This might suggest a role (yet unknown) of the testes in the control of male hydrocarbons.

5. Conclusion

This study demonstrates a role of sex peptide on male courtship behavior. Moreover, the data with DTA-E and RNAi knocked-down flies show the importance of the integrity of the testes (not the sperm) in the control of CHCs.

Acknowledgements

We want to thank Dr. Jacques Montagne for helpful suggestions on the manuscript. Funding was provided by the French Ministry of Research and Education.

Author details

Béatrice Denis, Benjamin Morel and Claude Wicker-Thomas*

*Address all correspondence to: claude.wicker-thomas@egce.cnrs-gif.fr

Laboratoire Evolution, Génomes, Comportements, Ecologie, UMR 9191, CNRS, IRD,
Université Paris-Sud and Université Paris-Saclay, Gif-sur-Yvette Cedex, France

References

[1] Findlay GD, Maccoss MJ, Swanson WJ. Mint: Proteomics reveals novel *Drosophila* seminal fluid proteins transferred at mating. PLoS Biology. 2008;**6**:e178. DOI: 10.1371/journal. pbio.0060178

[2] Avila FW, Sirot LK, LaFlamme BA, Rubinstein CD, Wolfner MF. Mint: Insect seminal fluid proteins: Identification and function. Annual Review of Entomology. 2011;**56**: 21-40. DOI: 10.1146/annurev-ento-120709-144823

[3] Chapman T, Bangham J, Vinti G, Seifried B, Lung O, Wolfner MF, Smith HK, Partridge L. Mint: The sex peptide of *Drosophila melanogaster*: Female post-mating responses analyzed by using RNA interference. Proceedings of the National Academy of Sciences. 2003;**100**:9923-9928. DOI: 10.1073/ pnas.1631635100

[4] Liu H, Kubli E. Mint: Sex-peptide is the molecular basis of the sperm effect in *Drosophila melanogaster*. Proceedings of the National Academy of Sciences. 2003;**100**:9929-9933. DOI: 10.1073/ pnas. 1631700100

[5] Soller M, Bownes M, Kubli E. Mint: Control of oocyte maturation in sexually mature *Drosophila* females. Developmental Biology. 1999;**208**:337-351. DOI: 10.1006/dbio.1999.9210

[6] Heifetz Y, Lung O, Frongillo EA Jr, Wolfner MF. Mint: The sex peptide of *Drosophila melanogaster*: The *Drosophila* seminal fluid protein Acp26Aa stimulates release of oocytes by the ovary. Current Biology. 2000;**10**:99-102

[7] Carvalho GB, Kapahi P, Anderson DJ, Benzer S. Mint: Allocrine modulation of feeding behavior by the sex peptide of *Drosophila*. Current Biology. 2006;**16**:692-696. DOI: 10.1016/j.cub.2006.02.064

[8] Ribeiro C, Dickson BJ. Mint: Sex peptide receptor and neuronal TOR/S6K signaling modulate nutrient balancing in *Drosophila*. Current Biology. 2010;**20**:1000-1005. DOI: 10.1016/j.cub.2010.03.061

[9] Cognigni P, Bailey AP, Miguel-Aliaga I. Mint: Enteric neurons and systemic signals couple nutritional and reproductive status with intestinal homeostasis. Cell Metabolism. 2011;**13**:92-104. DOI: 10.1016/j.cmet.2010.12.010

[10] Isaac RE, Li C, Leedale AE, Shirras AD. Mint: *Drosophila* male sex peptide inhibits siesta sleep and promotes locomotor activity in the post-mated female. Proceedings of the Biological Sciences. 2010;**277**:65-70. DOI: 10.1098/rspb.2009.1236

[11] Avila FW, Ravi Ram K, Bloch Qazi MC, Wolfner MF. Mint: Sex peptide is required for the efficient release of stored sperm in mated *Drosophila* females. Genetics. 2010;**186**: 595-600. DOI: 10.1534/genetics.110.119735

[12] Yapici N, Kim YJ, Ribeiro C, Dickson BJ. Mint: A receptor that mediates the post-mating switch in *Drosophila* reproductive behaviour. Nature. 2008;**451**:33-37. DOI: 10.1038/nature06483

[13] Häsemeyer M, Yapici N, Heberlein U, Dickson BJ. Mint: Sensory neurons in the *Drosophila* genital tract regulate female reproductive behavior. Neuron. 2009;**61**:511-518. DOI: 10.1016/j.neuron.2009.01.009

[14] Yang CH, Rumpf S, Xiang Y, Gordon MD, Song W, Jan LY, Jan YN. Mint: Control of the postmating behavioral switch in *Drosophila* females by internal sensory neurons. Neuron. 2009;**61**:519-526. DOI: 10.1016/j.neuron.2008.12.021

[15] Peng J, Zipperlen P, Kubli E. Mint: *Drosophila* sex-peptide stimulates female innate immune system after mating via the Toll and Imd pathways. Current Biology. 2005;**15**: 1690-1694. DOI: 10.1016/j.cub.2005.08.048

[16] Moshitzky P, Fleischmann I, Chaimov N, Saudan P, Klauser S, Kubli E, Applebaum SW. Mint: Sex-peptide activates juvenile hormone biosynthesis in the *Drosophila melanogaster* corpus allatum. Archives of Insect Biochemistry and Physiology. 1996;**32**:363-374. DOI: 10.1002/(SICI)1520-6327(1996)32:3/4<363::AID-ARCH9>3.0.CO;2-T

[17] Bontonou G, Shaik HA, Denis B, Wicker-Thomas C. Mint: Acp70A regulates *Drosophila* pheromones through juvenile hormone induction. Insect Biochemistry and Molecular Biology. 2015;**56**:36-49. DOI: 10.1016/j.ibmb.2014.11.008

[18] Kalb JM, DiBenedetto AJ, Wolfner MF. Mint: Probing the function of *Drosophila melanogaster* accessory glands by directed cell ablation. Proceedings of the National Academy of Sciences. 1993;**90**:8093-8097

[19] Robinow S, White K. Mint: Characterization and spatial distribution of the ELAV protein during *Drosophila melanogaster* development. Journal of Neurobiology. 1991;**22**:443-461

[20] Konrad KD, Marsh JL. Mint: Developmental expression and spatial distribution of dopa decarboxylase in *Drosophila*. Developmental Biology. 1987;**122**:172-185

[21] Ferveur JF, Savarit F, O'Kane CJ, Sureau G, Greenspan RJ, Jallon JM. Mint: Genetic feminization of pheromones and its behavioral consequences in *Drosophila* males. Science. 1997;**276**:1555-1558

[22] Billeter JC, Atallah J, Krupp JJ, Millar JG, Levine JD. Mint: Specialized cells tag sexual and species identity in *Drosophila melanogaster*. Nature. 2009;**461**:987-991. DOI: 10.1038/nature08495

[23] Harrison DA, Binari R, Nahreini TS, Gilman M, Perrimon N. Mint: Activation of a *Drosophila* Janus kinase (JAK) causes hematopoietic neoplasia and developmental defects. The EMBO Journal. 1995;**14**:2857-2865. DOI: 10.1038/nature08495

[24] Gutierrez E, Wiggins D, Fielding B, Gould AP. Mint: Specialized hepatocyte-like cells regulate *Drosophila* lipid metabolism. Nature. 2007;**445**:275-280. DOI: 10.1038/nature05382

[25] Parvy JP, Napal L, Rubin T, Poidevin M, Perrin L, Wicker-Thomas C, Montagne J. *Drosophila melanogaster* Acetyl-CoA-carboxylase sustains a fatty acid-dependent remote signal to waterproof the respiratory system. PLoS Genetics. 2012;**8**:e1002925. DOI: 10.1371/journal.pgen.1002925

[26] Giansanti MG, Farkas RM, Bonaccorsi S, Lindsley DL, Wakimoto BT, Fuller MT, Gatti M. Mint: Genetic dissection of meiotic cytokinesis in *Drosophila* males. Molecular Biology of the Cell. 2004;**15**:2509-2522. DOI: 10.1091/mbc.E03-08-0603

[27] Jung A, Hollmann M, Schafer MA. Mint the fatty acid elongase NOA is necessary for viability and has a somatic role in *Drosophila* sperm development. Journal of Cell Science. 2008;**120**:2924-2934. DOI: 10.1242/jcs.006551

[28] Szafer-Glusman E, Giansanti MG, Nishihama R, Bolival B, Pringle J, Gatti M, Fuller MT. Mint: A role for very-long-chain fatty acids in furrow ingression during cytokinesis in *Drosophila* spermatocytes. Current Biology. 2008;**18**:1426-1431. DOI: 10.1016/j.cub.2008.08.061

[29] DiBenedetto AJ, Harada HA, Wolfner MF. Mint: Structure, cell-specific expression, and mating-induced regulation of a *Drosophila melanogaster* male accessory gland gene. Developmental Biology. 1990;**139**:134-148

[30] Gligorov D, Sitnik JL, Maeda RK, Wolfner MF, Karch F. Mint: A novel function for the Hox gene Abd-B in the male accessory gland regulates the long-term female post-mating response in *Drosophila*. PLoS Genetics. 2013;**9**:e1003395. DOI: 10.1371/journal.pgen.1003395

www.ingramcontent.com/pod-product-compliance
Lightning Source LLC
Chambersburg PA
CBHW081235190326
41458CB00016B/5786